Textile Science and Clothing Technology

Series Editor

Subramanian Senthilkannan Muthu, Chief Sustainability Officer
Green Story, Canada
Tung Chung, Lantau, Hong Kong

This series aims to broadly cover all the aspects related to textiles science and technology and clothing science and technology. Below are the areas fall under the aims and scope of this series, but not limited to: Production and properties of various natural and synthetic fibres; Production and properties of different yarns, fabrics and apparels; Manufacturing aspects of textiles and clothing; Modelling and Simulation aspects related to textiles and clothing; Production and properties of Nonwovens; Evaluation/testing of various properties of textiles and clothing products; Supply chain management of textiles and clothing; Aspects related to Clothing Science such as comfort; Functional aspects and evaluation of textiles; Textile biomaterials and bioengineering; Nano, micro, smart, sport and intelligent textiles; Various aspects of industrial and technical applications of textiles and clothing; Apparel manufacturing and engineering; New developments and applications pertaining to textiles and clothing materials and their manufacturing methods; Textile design aspects; Sustainable fashion and textiles; Green Textiles and Eco-Fashion; Sustainability aspects of textiles and clothing; Environmental assessments of textiles and clothing supply chain; Green Composites; Sustainable Luxury and Sustainable Consumption; Waste Management in Textiles; Sustainability Standards and Green labels; Social and Economic Sustainability of Textiles and Clothing.

Subramanian Senthilkannan Muthu
Editor

Sustainable Nano Textile Technology

Case Studies

Springer

Editor
Subramanian Senthilkannan Muthu
Chief Sustainability Officer
Green Story, Canada
Tung Chung, Lantau, Hong Kong

ISSN 2197-9863 ISSN 2197-9871 (electronic)
Textile Science and Clothing Technology
ISBN 978-3-032-05901-7 ISBN 978-3-032-05902-4 (eBook)
https://doi.org/10.1007/978-3-032-05902-4

This Springer imprint is published by the registered company Springer Nature Switzerland AG
The registered company address is: Gewerbestrasse 11, 6330 Cham, Switzerland

This book is dedicated to:
The lotus feet of my beloved Lord Pazhaniandavar
My beloved late Father
My beloved Mother
My beloved Wife Karpagam and Daughters—Anu and Karthika
My beloved Brother
Last but not least
To everyone working in various industrial spaces to make the world "SUSTAINABLE"

Contents

Sustainable Nanotextile Technology: Case Studies

Subramanian Senthilkannan Muthu

The textile industry, a cornerstone of global manufacturing, faces mounting pressure to reconcile its economic significance with environmental stewardship. Traditional textile processes are often resource-intensive, contributing substantially to water pollution, energy consumption, and waste generation. In response, the integration of nanotechnology presents a transformative pathway toward sustainability. By manipulating materials at the nanoscale, textiles can be endowed with enhanced properties such as increased durability, water and stain resistance, and antimicrobial functions, all while potentially reducing the need for harmful chemicals and excessive water use. Moreover, advancements in green nanotechnology emphasize the development of eco-friendly nanomaterials and processes, aligning with principles of green chemistry and engineering. This convergence of nanoscience and textile manufacturing not only addresses environmental concerns but also opens avenues for innovation in smart textiles and wearable technologies. As the demand for sustainable solutions intensifies, exploring the role of nanotechnology in revolutionizing textile production becomes imperative. This book delves into the principles, applications, and future prospects of sustainable nanotextile technology, aiming to provide a comprehensive understanding of its potential to reshape the textile industry's environmental footprint.

The second chapter deals with enhancing the functionality of natural fiber textiles through in situ nanofinishing techniques: A sustainable and cost-effective approach. The integration of sustainable practices in textile production has become important in the pursuit for eco-friendly solutions in the industry. This chapter explores in situ nanofinishing techniques for natural fiber-based textiles, focusing on enhancing their performance and functionality while minimizing environmental impact. In situ nanofinishing involves the synthesis and deposition of nanomaterials directly onto textile substrates during processing, offering advantages such as

S. S. Muthu (✉)
Chief Sustainability Officer, Green Story Inc., Toronto, Canada

S. S. Muthu (ed.), *Sustainable Nano Textile Technology*, Textile Science and Clothing Technology, https://doi.org/10.1007/978-3-032-05902-4_1

reduced chemical usage, energy efficiency, and waste minimization. Various in situ nanofinishing approaches for natural fibers such as cotton, wool, jute, and hemp, highlighting their potential to impart properties such as antimicrobial activity, UV protection, water repellency, and enhanced durability are discussed in this chapter. Though nanosilver is widely explored for application in textiles, other nanomaterials like copper, iron oxide, zinc oxide, titania, chitosan, triclosan are also being explored for diverse functionalities. The role of biodegradable and green nanomaterials in achieving these finishes is discussed, emphasizing the alignment of this technology with sustainable development goals. Additionally, case studies are presented to demonstrate the effectiveness of in situ nanofinishing in maintaining the inherent qualities of natural fiber while adding value through functional enhancements. This chapter also addresses the scalability and commercialization potential of this technology, along with its challenges, including cost factors and regulatory considerations.

The third chapter is about the Sustainable Nanotextiles—A review on innovations and Challenges. Nanotechnology is an emerging field covering a range of scientific and technological categories, moving forward in the textile industry. For instance, the ability to manipulate materials at nanoscale one billionth of a meter has led to the fabrication of textiles with desirable features like stain resistance, wrinkle-free properties, static-eliminating properties, and electrical conductivity while maintaining comfort and flexibility. This article explores the impact of nanotechnology on textiles, especially sustainable innovations. Nanomaterials can also be integrated into fabrics to develop multifunctional fabrics, such as antimicrobial, UV resistant, flame-retardant, and electrically conductive, leading to more durable and sustainable products. In addition, smart nanomaterial-based devices enable textiles to react to external factors such as a change in electrical signal, color, or physiological state, contributing to the development of connected clothes with potential applications across different sectors such as fashion, healthcare, defense, and sports. The review notes the growing demand for high-performance sustainable textiles and give insights on the approaches to integrate nanomaterials affordably and sustainably. It also discusses the potential health and environmental risks associated with nanotextiles, including nanotoxicity and their ecological consequences as well as the importance of conducting life cycle assessments (LCA) of new nanotechnology and developing green nanotechnology. The current problems in developing green and biocompatible nanomaterials along other potential applications of smart textiles in energy harvesting, environmental monitoring, sustainability is also highlighted. In a timely context, this chapter discusses current trends and challenges focusing on functionalizing textiles with nanomaterials to be in line with societal expectations of safety, cost-effectiveness, and sustainability alongside addressing potential health and environmental issues. The future of smart textiles utilizing nanomaterials for energy conversion and environmental monitoring is also discussed alongside effective waste disposal practices, emphasizing green nanotechnology‘s ability to avoid harmful effects. Finally, it highlights the challenges of integrating such green nanomaterials in these innovations and the need for further research toward greener environment emphasizing the importance of advancing textiles safely with limited adverse effects.

The fourth chapter is about the Nanotextiles for Sustainability and Circular Economy. This chapter "Nanotextiles for Sustainability and Circular Economy" delves into how nanotechnology has been bringing about a significant change in the paradigms of sustainability in the textile industry which are also in tune with the principles of the circular economy. In addition, it presents models for environmentally conscious production and discusses future trends in nanotextile waste. This chapter discusses the benefits of circular design approaches—closed-loop production and upcycling of materials—to reduce waste and increase the use of renewable resources. Examples of case studies demonstrate the translation of nano-enabling sustainable solutions in industries and how regulatory issues and consumer acceptability on these novel solutions are pursued. It is envisaged in this chapter that while these technologies remain innovative and that, while sustainable practices are developed, in the future itself, nanotextiles would bring the shift to the regenerative and eco-friendly textile industry.

The fifth chapter deals with the advancing textiles with nanotechnology: Innovations, applications, and sustainability trends. Sustainable Nanotextile technology represents a ground breaking advancement in the textile industry, merging nanotechnology with eco-friendly practices to create high-performance materials with minimal environmental impact. Nanotextile products are innovative materials enhanced with nanotechnology to deliver advanced functionalities beyond traditional textiles. The objectives of nanotechnology encompass a wide range of scientific, industrial, and societal goals, aimed at leveraging the unique properties of nanoscale materials and structures. The products are designed for various industries and offer improved performance, durability, and sustainability. This interdisciplinary approach leverages nanoscale engineering to enhance textile properties such as durability, water repellency, antibacterial activity, UV protection, and self-cleaning capabilities, all while promoting resource efficiency. This chapter also analyzes the consolidation of nanotechnology in the textiles market to assess worldwide trends and patent coverage. Here are some notable nanotechnology textile products and their applications that are collectively taken for discussion, such as antimicrobial products for medical and surgical products; self-cleaning for laboratory uniforms; water-repellent for sports/home textile products; UV-resistant fabrics for sportswear; lightweight and high-strength fabrics for military wear.

The sixth chapter deals with the sustainability and functional performance of nanoscience in textiles. The use of nanotechnology in the apparel industry is not a recent development. Many apparel firms began adding silver nanoparticles to their goods in the middle of the 2000s. Because silver nanoparticles are antibacterial, they eliminate the microorganisms that produce unpleasant odors. The finished clothing requires fewer washings as these nanoparticles are added to the fabric to stop odor. These nano-infused textiles are still in demand today and come in various techniques, from t-shirts to socks. The field of nanofabrics is producing far more advanced technologies than only antibacterial nanoparticles. Self-cleaning textiles, water-repellent textiles, and clothing that may lessen odors by chemically altering chemical compounds that create unpleasant odors are further desirable clothing qualities that might be accomplished with nanotechnology. These developments would benefit from the unique characteristics of nanomaterials, especially their high surface area-to-volume ratio, which increases the exposure of active surfaces to the environment. It is possible that more recent nanofabric technologies have significant issues of their own. However, it is worthwhile to investigate the possible advantages of using textiles strengthened with nanotechnology. And this area of nanotechnology can only expand as long as science keeps improving and enabling us to solve environmental issues. However, several textile materials up to the nanometre scale have been successfully produced using nanotechnology and nanoscience, having extremely porous fibers and large surface area that demonstrate their viability for advanced applications. Because of their unique structures and nanoscopic size, these materials have exceptional capabilities that provide distinctive electrical, magnetic, optical, biological, thermal, and mechanical features. The nanofibers can be created using various methods, such as phase separation, centrifugal spinning, electrospinning, melt blowing, drawing, template synthesis, and biocomponent extrusion. The possible applications of nanofibers include tissue engineering, thermal-resistant materials, biomedical products, filtration down to the nanometre scale, breathable sportswear, and protective apparel for military, chemical, and medical personnel. These cutting-edge application textiles may be produced effectively by using nanotechnology and reinforcing the textiles with nanofiber, sand, and nanoparticle additions. A notable development at present is the incorporation of environmental-friendly nanomaterials, such biodegradable and bio-based nanoparticles, which promote more sustainable production methods. These innovations improve textile performance by providing qualities including water resistance, UV protection, and self-cleaning properties, in addition to addressing environmental concerns. It is anticipated that future developments will focus on improving the synthesis of nanomaterials, increasing production, and maintaining the sustainability of the entire lifespan. It is projected that new developments like smart features and circular economy strategies would further transform this industry.

I take this opportunity to thank all the contributors for their earnest efforts to bring out this book successfully. I am sure readers of this book will find it very useful.

With best wishes,
Subramanian Senthilkannan Muthu

Enhancing the Functionality of Natural Fiber Textiles Through In Situ Nanofinishing Techniques: A Sustainable and Cost-Effective Approach

N. Vigneshwaran and A. Arputharaj

Abstract The integration of sustainable practices in textile production has become important in the pursuit for eco-friendly solutions in the industry. This chapter explores in situ nanofinishing techniques for natural fiber-based textiles, focusing on enhancing their performance and functionality while minimizing environmental impact. In situ nanofinishing involves the synthesis and deposition of nanomaterials directly onto textile substrates during processing, offering advantages such as reduced chemical usage, energy efficiency, and waste minimization. Various in situ nanofinishing approaches for natural fibers like cotton, wool, jute, and hemp, highlighting their potential to impart properties such as antimicrobial activity, UV protection, water repellency, and enhanced durability are discussed in this chapter. Though nanosilver is widely explored for application in textiles, other nanomaterials like copper, iron oxide, zinc oxide, titania, chitosan, triclosan are also being explored for diverse functionalities. The role of biodegradable and green nanomaterials in achieving these finishes is discussed, emphasizing the alignment of this technology with sustainable development goals. Additionally, case studies are presented to demonstrate the effectiveness of in situ nanofinishing in maintaining the inherent qualities of natural fiber while adding value through functional enhancements. This chapter also addresses the scalability and commercialization potential of this technology, along with its challenges, including cost factors and regulatory considerations.

Keywords Nanofinishing · Natural fibers · In situ · Nanotoxicity · Antimicrobial textiles · UV protection · Sustainable textiles

N. Vigneshwaran · A. Arputharaj (✉)
Chemical and biochemical Processing Division, ICAR-Central Institute for Research on Cotton Technology (ICAR-CIRCOT), Mumbai, India

S. S. Muthu (ed.), *Sustainable Nano Textile Technology*, Textile Science and Clothing Technology, https://doi.org/10.1007/978-3-032-05902-4_2

1 Introduction

The advent of nanoscience and nanotechnology has revolutionized textile finishing, introducing an advanced domain known as nanofinishing. This cutting-edge approach enables the enhancement of textile materials with various functional properties, making them more durable, efficient, and versatile. The commercial potential of nanotechnology in the textile industry is immense, primarily because conventional methods of fabric treatment often result in effects that are temporary and diminish over time due to repeated laundering or wear. In contrast, nanotechnology offers a transformative solution by imparting long-lasting functionalities to textiles. Nanoparticles (NPs) exhibit a remarkably high surface area-to-volume ratio and possess significant surface energy, which enhances their interaction with fabrics. This results in an improved affinity between the nanoparticles and textile fibers, leading to increased durability of the functional properties. Additionally, the application of nanoparticles does not compromise the breathability, softness, or overall hand feel of fabrics, making them an ideal choice for textile enhancements. Consequently, the integration of nanotechnology into the textile sector has garnered significant attention and continues to gain momentum as a promising frontier in material science. Although the textile industry represents a relatively small segment of global research in nanotechnology, it has been among the pioneers in successfully implementing these advancements. The fibers and textiles industries were among the first to explore and commercialize nanotechnology applications, demonstrating its feasibility and benefits for consumer usage. Today, various nanoengineered materials have been developed to impart superior functionalities to textiles, enhancing their performance in diverse applications. Among the myriad of nanomaterials employed in textile finishing, nano-sized forms of zinc oxide (ZnO), titanium dioxide (TiO_2), silver (Ag), and copper oxide (CuO) have gained prominence. These materials exhibit exceptional antibacterial, ultraviolet (UV) absorption, and photocatalytic properties, making them highly desirable for textile applications. Each of these nanomaterials contributes unique advantages: silver nanoparticles provide potent antimicrobial effects, titanium dioxide offers UV protection and self-cleaning property. Of these, nano-ZnO has emerged as a frontrunner due to its distinct attributes. Its chemical inertness ensures stability, its intrinsic whiteness makes it aesthetically suitable for textiles, and its cost-effectiveness renders it an economically viable choice for large-scale applications. The integration of nano-ZnO in textiles not only enhances antibacterial resistance but also improves UV shielding, making it an indispensable component in the development of advanced functional fabrics. There are different application techniques of NPs on textile materials. Application of NPs to textiles is primarily carried out by means of wet-chemical procedures using synthetized or ready-made NPs. Synthetized NPs are applied to cotton fabrics by impregnating with a wet pick-up, drying, and curing method.

This chapter introduces the growing importance of sustainability in textile production and the need for eco-friendly alternatives. It then explores into in situ nanofinishing techniques, explaining their process, advantages, and suitability for natural fiber-based textiles. Various nanomaterials such as silver, zinc oxide, chitosan, and others are explored for their ability to impart functional properties like antimicrobial activity, UV protection, and water repellency. The discussion extends to the role of biodegradable nanomaterials and their alignment with sustainable development goals, supported by case studies demonstrating practical applications. Finally, the chapter addresses the scalability and commercialization of in situ nanofinishing, along with associated challenges such as cost, regulations, and market readiness.

1.1 Pad-Dry-Cure Method

A padding mangle is used for the purpose of applying the nanoparticles from the suspension to the fabric surface. When the fabric is fed to the nip of the padding rollers, the pressure of the nip helps the NPs to be pressed on to the fabric surface (Fig. 1). The treated fabric is then dried and then cured at a high temperature to ensure the fixation of NPs onto fabric surface.

This method invariably uses a binder chemical for binding of nanoparticles onto fabric surface. It is important to check the compatibility of binder with that of dispersing agent and nanoparticles. This method also requires higher quantity of NPs and liquor as it necessitates the use of a padding mangle.

In conventional finishing techniques, pad-dry-cure, thermal, radiation, or chemical methods are generally employed to fix these nanoparticles on textiles. The process sequence in such cases either involves several steps or is chemically intensive. The coatings produced are thick, unstable, and do not allow the fabric to breathe properly. Our research group [30] applied 2% (owm) ZnO-soluble starch nanocomposites synthesized using water as a solvent and soluble starch as a stabilizer onto

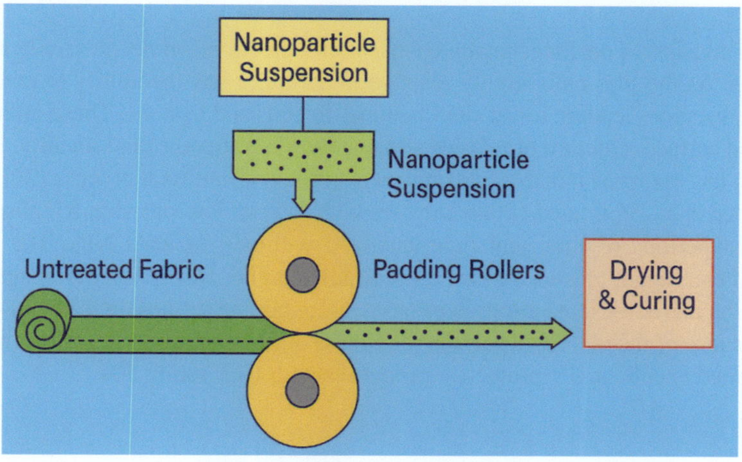

Fig. 1 Padding method of application of NPs on textiles

cotton fabrics to impart antibacterial and UV protection functions by pad-dry-cure method using acrylic binder. Salama and El-Sayed [22] used polymeric binders namely, poly vinyl pyrrolidone, poly-4-vinyl pyridine, and reactive polyurethane resin to increase the binding properties of nano-ZnO particles on cellulosic fabric. The coating was performed by pad-batch-method and the coated fabrics was assessed for their antimicrobial properties against *Escherichia coli* and *Staphylococcus aureus*. They reported higher concentrations of the polymeric binder used for the coating lead to a decrease in the bacterial reduction and could be attributed to the aggregation of the nano-ZnO on the polymers surface, losing the advantage of the nanosize.

1.2 Layer-by-Layer Method (LbL)

LbL method can provide a chemically mild alternative to these processes for producing ultrathin, transmissive, and stable coatings. Uğur et al. [28] developed ZnO NP-based multilayer nanocomposite films on cationized woven cotton fabrics via layer-by-layer molecular self-assembly technique. For cationic surface charge, cotton fabrics were pre-treated with 2,3-epoxypropyltrimethylammonium chloride (EP3MAC) by pad-batch method. XPS and SEM were used to examine the deposited nano-ZnO multilayer films on the cotton fabrics. Nano-ZnO films deposited on cotton fabrics exhibited excellent antimicrobial activity against *S. aureus* bacteria. The results also showed that nano-ZnO multilayer films enhanced the protection of cotton fabrics from UV radiation.

1.3 Plasma Pre-treatment

Plasma consists of positive, negative ions, atoms, electrons, molecules, radicals, and photons. As the plasma is highly reactive in nature, it has the ability to react with other substances leading to various chemical fusions and fissions. These effects can alter the surface structure of textiles and resulting in different functionality. Plasma process has many sustainable advantages over the conventional process [24].

Jazbec et al. [9] treated cotton samples with oxygen low-pressure RF plasma for different periods of time and functionalized with 3% of ZnO NPs. The results showed that longer plasma treatment time causes higher adsorption of nanoparticles due to the increased concentration of oxygen-containing groups on the fiber surface and because of the increased surface roughness of fibers. Plasma-treated cotton fabrics gained excellent UV protective properties with UPF rating 50+.

1.4 Electrospraying Method

Electrospraying is a well-known technique of liquid atomization employing electrical forces for controlled application of chemicals. Vitchuli et al. [31] prepared ZnO/Nylon 6 nanofiber mats by an electrospinning–electrospraying hybrid process in which ZnO NPs were dispersed on the surface of Nylon 6 nanofibers without becoming completely embedded. The prepared ZnO/Nylon 6 nanofiber mats were evaluated for their abilities to kill bacteria or inhibit their growth and to catalytically detoxify chemicals. Results showed that these ZnO/Nylon 6 nanofiber mats had excellent antibacterial efficiency (99.99%) against both the Gram-negative *Escherichia coli* and Gram-positive *Bacillus cereus* bacteria.

1.5 Atomic Layer Deposition Method

Atomic layer deposition (ALD) is a vapor phase technique capable of producing thin films of variety of materials. Jur et al. [11] coated ZnO by ALD method on cotton fabric and paper to increase the conductivity.

2 In Situ Nanofinishing of Textiles

The primary problem with fixing nanoparticles on textiles is achieving uniform and durable adhesion of the nanoparticles to the cotton fibers, which can be challenging due to the natural hydrophilic nature of cotton, potentially leading to uneven distribution and wash-induced loss of nanoparticles from the fabric over time. This can significantly impact the functionality of the treated textile, such as its antimicrobial or stain-resistant properties. The advancement of nanofinishing in textiles has led to the development of in situ nanofinishing, a technique that enhances the functional properties of cotton fabrics at the molecular level. This process involves the formation of nanoparticles directly on the textile surface without the need for external nanoparticle synthesis and post-application.

Compared to conventional finishing methods, in situ nanofinishing ensures better adhesion, durability, and uniform dispersion of nanoparticles, making it a promising approach for producing high-performance textiles. In situ nanofinishing can overcome these drawbacks by incorporating functional nanoparticles directly onto textile fibers, thereby imparting properties such as antibacterial activity, UV protection, water repellency, and flame retardancy properties. In situ nanofinishing is typically carried out using sol–gel techniques, chemical reduction, or enzymatic synthesis, where the nanoparticles are generated directly within the textile matrix (Fig. 2).

The process involves immersing cotton fabrics in precursor solutions containing metal salts or other nanomaterial precursors. Upon chemical treatment or exposure

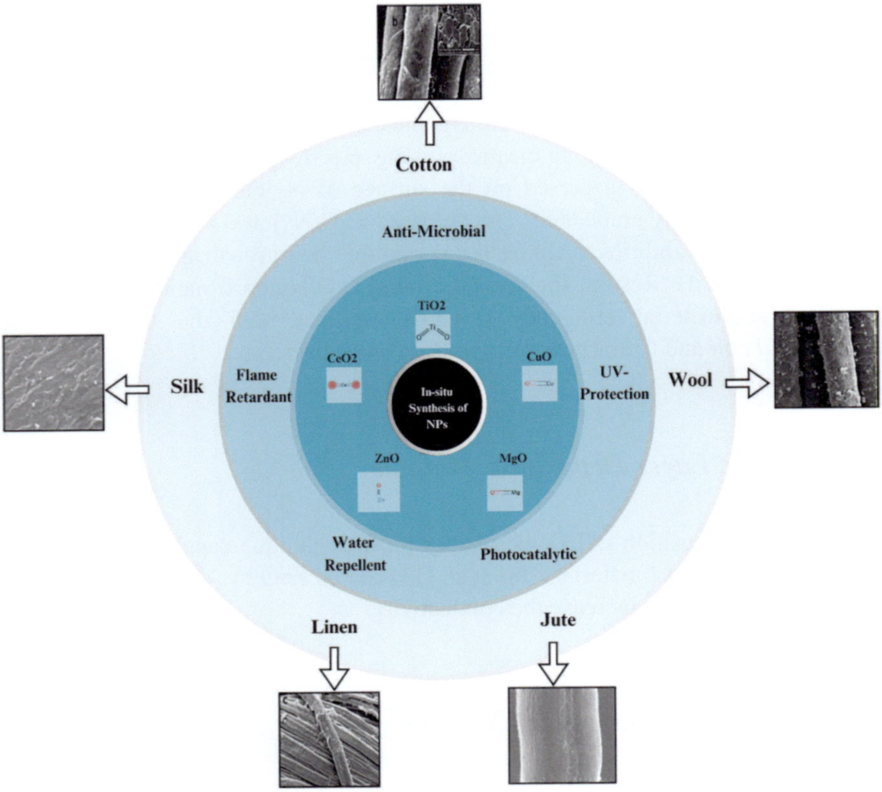

Fig. 2 In situ nanofinishing of various natural textiles

to heat, the nanoparticles are formed in situ and strongly adhere to the fiber surface, ensuring long-lasting functionality.

2.1 In Situ Finishing of Cotton Using Nano-Ag

A commonly used approach is in situ synthesis of silver nanoparticles (nano-Ag) on cotton textiles, where silver nitrate ($AgNO_3$) solution is applied to the fabric, followed by a reducing agent (such as glucose, sodium borohydride, or plant extracts). The reduction reaction facilitates the formation of silver nanoparticles, which are embedded onto the cotton fibers, imparting antimicrobial and odor-resistant properties to the fabric.

Similar technique is applicable for all types of metallic nanoparticles. Nano-Ag, when deposited on the surface of cotton textiles, impart both color (due to surface plasmon resonance) and antibacterial property. The in situ formation of nano-Ag on cotton fibers is based on a bottom-up approach, where silver ions are reduced to silver nanoparticles directly on the fiber surface. This process typically involves impregnating the cotton with a silver precursor like silver nitrate followed by a reduction reaction that facilitates the nucleation and growth of nanoparticles. Cotton fibers, composed mainly of cellulose, provide a natural template for nanoparticle formation due to the presence of hydroxyl(OH) groups, which can interact with silver ions and assist in nanoparticle stabilization. The uniform distribution of silver ions within the cotton matrix is crucial for achieving a well-dispersed nanoparticle coating with desirable functional properties.

The reduction of silver ions (Ag^+) to metallic silver (Ag^0) is a critical step in nanoparticle formation. This can be achieved through chemical, biological, or physical reduction methods. In chemical methods, reducing agents such as sodium borohydride ($NaBH_4$), glucose, or ascorbic acid donate electrons to Ag^+, leading to the formation of metallic silver nanoparticles. In biologically assisted synthesis, plant extracts, bacterial metabolites, or even the cellulose itself can act as mild reducing and stabilizing agents. The in situ approach ensures that nanoparticles are generated and deposited directly onto the fiber surface, eliminating the need for additional functionalization steps and improving durability through strong interactions between the nanoparticles and cellulose matrix. The use of reducing agent like is reported earlier, wherein the cotton fabric is first treated with caustic soda to activate the OH groups on the cellulose surface and then the nano-Ag is introduced on fabric by the reduction of $AgNO_3$ salt using ascorbic acid [26]. The chemical reaction is given in Eq. 1. The incorporation of nano-Ag increases of dyestuff exhaustion (7%) and color strength (15%) of the fabric in addition to exhibiting more than 90% antibacterial activity against both Gram-positive (*S. aureus*) and Gram-negative (*E. coli*) bacteria.

$$2Ag^{+} + C_6H_8O_6 \rightarrow 2Ag^{0} + C_6H_6O_6 + 2H^{+} \tag{1}$$

Once formed, nano-Ag exhibit unique antimicrobial and optical properties, making nano-Ag coated cotton highly valuable in applications such as antibacterial textiles, wound dressings, and smart fabrics. The size and distribution of nanoparticles can be controlled by adjusting synthesis parameters like precursor concentration, reaction time, temperature, and pH. Additionally, the stability of nano-Ag is enhanced by the natural capping effect provided by cellulose chains, which prevents aggregation. This method of in situ synthesis ensures cost-effective, scalable, and eco-friendly production of functional textiles with long-lasting antimicrobial and conductive properties. A sustainable inkjet printing technique was earlier demonstrated [5] to create antimicrobial cotton fabrics by a novel in situ process using bio-synthesized nano-Ag. The silver salt was used to form nanosilver by inkjet printing process. Silver salt loaded ink was used in this process and the substrate (cotton fabric) was pre-treated with carbohydrate polymers like sodium alginate. While printing, these carbohydrates helped to reduce the silver salts in situ, forming nanosilver inside the cotton structure. Thus produced cotton fabrics showed very good antibacterial activity. In an another study [14], silver nitrate was transformed to $[Ag(NH_3)_2]^+$ complex and then the nanosilver were loaded directly on cotton using the inherent reducing and stabilizing properties of cellulosic chains of the cotton fabric. The treated fabrics showed excellent antibacterial property.

2.2 *In Situ Finishing of Cotton Using Nano-ZnO*

In this process, cotton fabric is treated with a zinc precursor (such as zinc acetate or zinc nitrate) and an alkaline solution, leading to the formation of ZnO nanoparticles in situ. This technique enhances the UV protection capability of textiles in addition to its antibacterial property. Similar technique is applicable for all types of metal oxide nanoparticles. In our earlier research work [18], in situ synthesis of nano-ZnO onto 100% cotton fabrics (terry or woven) by spraying or dipping process resulted in durable antibacterial and UV protection finishes. The nano-sized pore structure of cellulosic fibrils acted as nucleation site for the formation of nano-ZnO from the precursors, zinc nitrate, and sodium hydroxide (Fig. 3). Mao et al. [12] reported in situ growth of ZnO on SiO_2 sol coated cotton fabrics via hydrothermal method using relatively higher reagent concentrations. The cotton fabric was then treated in hot water to obtain needle-shaped nano-ZnO crystallites. The UV protection performance of cotton fabric was studied. After water treatment at 100 °C or higher, the cotton fabric was covered with approximately 24 nm diameter needle-shaped ZnO. It was found that hot water treatment could transform the morphology of ZnO NPs on the surface of cotton fabric from sphere and rod to needle shape through a re-crystallization process.

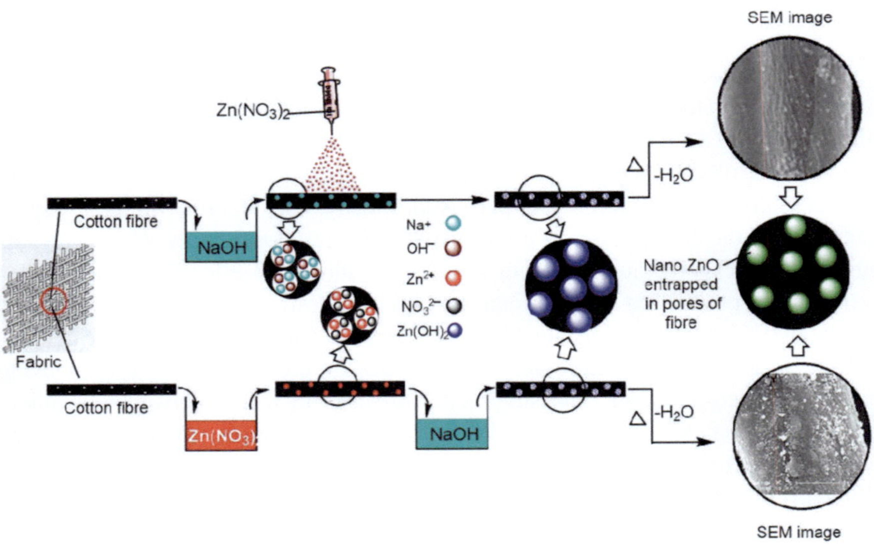

Fig. 3 In situ synthesis of nano-ZnO by spraying and dipping processes. ([18] Reproduced with permission from Elsevier)

Though hot water treatment could not significantly increase the amount of ZnO on the surface of cotton fabric, longer treatment time and higher treatment temperature can make the needle-shaped ZnO nano-rod have smaller diameter and better crystalline perfection. As a result, ZnO-coated cotton fabric has better UV-blocking property. After treating in boiling water for 3 h, ZnO-coated cotton fabric had excellent UV protection (UPF > 50). After 20 washes, cotton fabric still had good UV protection.

Perelshtein et al. [17] synthesized and deposited in situ ZnO NPs on surface of cotton fabrics using ultrasound irradiation. Optimization of process resulted in a homogeneous distribution of ZnO nanocrystals, 30 nm in size, on the fabric surface. The mechanism of ultrasound-assisted coating was proposed. The antibacterial activities of ZnO-fabric composite were tested against *Escherichia coli* and *Staphylococcus aureus*. A significant bactericidal effect, even in a 0.75 wt % coated fabric was demonstrated. Shateri-Khalilabad et al. [23] synthesized nano-ZnO by in situ method and applied on cotton fabric. The antimicrobial activity was measured against *Klebsiella pneumonia* and *Staphylococcus aureus* bacteria. Laundering durability was performed to evaluate the stability of antibacterial properties of the fabrics in analogy to the ISO 105-C02:1989(E) method. Nearly the same inhibition zone was observed for the samples in the dark and samples exposed to UV light. Results showed the effectiveness of synthesized ZnO particles in incidence of antimicrobial effect, even without the presence of light. It was found that the ZnO-coated fabrics had better antibacterial effect against *S. aureus* than *K. pneumonia*. The ZnO-coated fabric preserved very good durability in regard to antibacterial activity, even after five cycles of washing.

Another study [8] focused on in situ growth of nano-ZnO on 100% cotton fabrics by one-step hydrothermal method that imparted UV protecting, antibacterial, and photo-catalytic properties. Sodium hydroxide and Zinc nitrate hexahydrate were used as reactants for the growth of nano-ZnO on the cotton fabrics. Briefly, the solution of NaOH was added dropwise to 100 mL solution of $Zn(NO_3)_2{\cdot}6H_2O$. After 5 min of stirring, the resultant solution was transferred into the autoclave along with the cotton fabric and heated at 90 °C for 1 h followed by drying in the oven at 90 °C for 2 h. The mechanism of the formation of nano-ZnO is given in Eqs. 2–5.

$$\mathrm{Zn}\left(\mathrm{NO_3}\right)_2 \cdot 6\mathrm{H_2O} \rightarrow \mathrm{Zn}^{2+} + 2\mathrm{NO_3}^{-} + 6\mathrm{H_2O} \tag{2}$$

$$\mathrm{NaOH} \rightarrow \mathrm{Na}^{+} + \mathrm{OH}^{-} \tag{3}$$

$$\mathrm{Zn}^{2+} + 4\mathrm{OH}^{-} \rightarrow \mathrm{Zn}\left(\mathrm{OH}\right)_4^{\;2-} \tag{4}$$

$$\mathrm{Zn}\left(\mathrm{OH}\right)_4^{\;2-} \rightarrow \mathrm{ZnO} + \mathrm{H_2O} + 2\mathrm{OH}^{-} \tag{5}$$

In a novel green process [29], nano-ZnO in situ synthesis on cotton fabric was reported where zinc acetate was precipitated with water extracts from biowaste, such as pomegranate peel and wood ash. The results showed that in the synthesis process, both water extract of pomegranate peels along with an alkali source (wood ash) were required for the formation nano-ZnO from zinc acetate.

2.3 In Situ Finishing of Cotton Using Nano-Titania

Nano-titania are reported to have very good self-cleaning and UV-absorbing properties suitable for textile applications. For in situ synthesis, nano-titania was synthesized and loaded to the cotton fabrics simultaneously by hydrolysis of titanium tetra isopropoxide in the presence of water, acetic acid as a dispersant under ultrasonic irradiation [21]. The results from this study confirmed the formation of anatase titania nanoparticles with 3–6-nm crystalline size, loaded onto the cotton fabric that led to very good self-cleaning and UV protection properties.

2.4 In Situ Finishing of Cotton Using Other Nanomaterials

For finishing cotton with nano-CuO, the exhaust dyeing method was employed with four different stages [20]. In the exhaustion stage, the Cu^{2+} is moved from the exhaustion bath to the cotton textiles due to the decomposition of the copper acetate salt in aqueous media. In the fixation stage, a chemical reaction occurs between Cu^{2+} ions and hydroxide ions resulting from the complete ionization of NaOH in water. Here, copper hydroxide is formed and then transformed to CuO. In the neutralization stage, glacial acetic acid is used to neutralize the cotton textile and in rinse-off stage, the functionalized textiles are washed with water to remove the residues of acetic acid. The functionalized cotton textiles showed bacterial reduction against *Escherichia coli* between 89.7% and 99.7% and also an improvement in the UPF from 7 to 32. Another research reported the finishing of nano-gold on cotton by in situ process [25]. Here, the surface plasmon resonance of nano-gold imparted color to cotton fabrics that shows good colorfastness to washing and rubbing. This cotton treated with nano-gold exhibited significant catalytic activity for reduction of 4-nitrophenol by sodium borohydride and could be reused in this reaction. Also, the treatment with nano-gold improved the UV-blocking ability of the fabric and also its antibacterial activity.

A cotton fabric/titania/nanocellulose flexible and recyclable composite material was demonstrated with photocatalytic degradation activity and antibacterial property [7]. During the preparation, nano-titania were synthesized in situ on cotton fiber and wrapped with cellulose nanocrystals. The prepared composite showed excellent photocatalytic efficiency under weak UV irradiation. Also, it showed excellent antibacterial activity against *S. aureus* and *E. coli*. The schematic representation of nano-titania and nanocellulose deposition on cotton fabric is shown in Fig. 4. The elemental analysis of composite material is shown in Fig. 5.

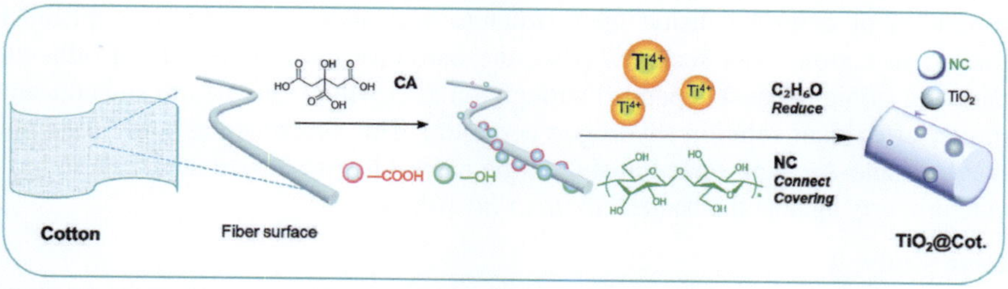

Fig. 4 Schematic representation of nano-titania and nanocellulose deposition on cotton fabric. [7] *Reproduced with permission from Royal Society of Chemistry*

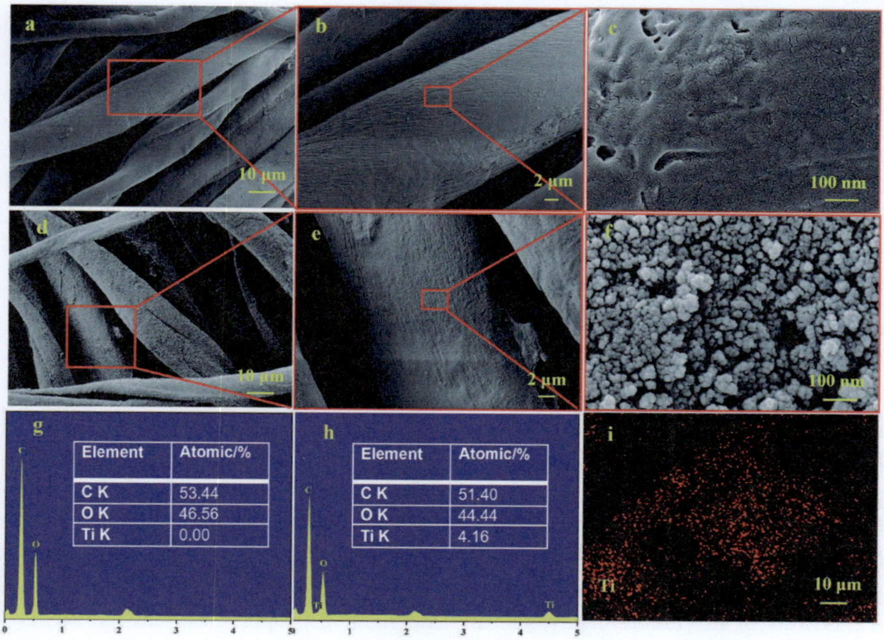

Fig. 5 Microstructure and elemental compositions of control and treated cotton fabrics. SEM images of control cotton (**a–c**); SEM images of composite cotton (**d–f**); EDS analysis of cotton (**g**); EDS analysis of composite cotton (**h**); and, EDS mapping of composite cotton (**i**). ([7]) Reproduced with permission from Royal Society of Chemistry)

Case Study: 1

The in situ synthesis of nano-Ag in five different textile fibers (cotton—untreated and chemically bleached, wool, polyamide, and polyester) was prepared and characterized a team of researchers [15, 16]. In this work, fibers were immersed silver ion solution (1 g/L of $AgNO_3$) at room temperature for 24 h followed by draining the solution and re-immersion in a strong chemical reducing solution (0.25 g/L of $NaBH_4$) for 24 h. This resulted in the formation of nano-Ag with the size ranging from 5 to 50 nm as deduced from FESEM images. Moreover, for natural fibers, the formation of nano-Ag inside their structure was also confirmed from FESEM images in cotton cross sections. Also, the oxidative dissolution rate of adhered nano-Ag (ionic release) depended strongly on fiber nature, and nature and concentration of oxidant solution (hydrogen peroxide). This fundamental study suggests that the nano-Ag synthesized in situ on the natural fibers showed an excellent biocide property against the bacterium, *Bacillus subtilis*.

Case Study: 2

Another research group [32] reported the production of flame-retardant fabric using phytic acid and in situ synthesized nano-Ag. The in situ synthesis process minimized the aggregation of nano-Ag. Also, the finished fabric exhibited both flame-retardant and antibacterial properties. The prepared textiles were evaluated for their combustion performance, thermal performance, and antibacterial capabilities and their microstructures were studied using XRD, SEM, and XPS. Their findings indicated excellent launderability [limiting oxygen index (LOI) = 31% after 20 washing cycles] for the flame-retardant property. The nano-Ag loaded flame-retardant textiles demonstrated self-extinguishing property with a LOI value of 27%. Also, the flame-retardant antibacterial textiles showed the clearing zone of inhibition against *Escherichia coli* and *Staphylococcus aureus,* 5.28 mm and 4.32 mm, respectively. In situ synthesis of nano-Ag loaded flame-retardant textiles successfully reduced agglomeration and heterogeneous dispersion of nanomaterials.

Case Study: 3

Our research group earlier reported [4] the development of a durable multifunctional finish for cotton fabric by in situ synthesis of nano-ZnO. Here, zinc nitrate interacted with cellulose molecules and opened up the internal pore structure of fibrils that in turn forms nucleating sites for nano-ZnO growth, on reaction with sodium hydroxide. The in situ nano-ZnO finished fabric showed excellent antibacterial activities (>99%) against two pathogens, *Staphylococcus aureus* and *Klebsiella pneumoniae* and very good UV protection factor (40 UPF). Also, these properties were not significantly affected even after 30 washes. The in situ synthesis process did not affect the inherent physical and mechanical properties of the fabric. The total hand value of the finished fabric changed from 3.54 to 3.40 as analyzed by the Kawabata Evaluation System.

Case Study: 4

Another research work [6] explored the antibacterial and UV protection properties of cotton fabric having in situ-synthesized copper oxide-titanium dioxide nanocomposite particles. Here, cotton fabrics treated with L-methionine were functionalized with bimetallic nanocomposite particles by pad and rolling technique. Copper nitrate and titanium isopropoxide were used as precursors for synthesis. The method yielded a crystallite size of 18.09 and 32.5 nm, copper oxide and titania, respectively. The results of various analyses showed that an esterification reaction occurred between L-methionine and OH group in cotton fabric that further enhanced the binding of nanocomposite particles. The treated fabrics exhibited excellent antibacterial activity against *S. aureus* and *E. coli*; and, UPF of 50+.

Case Study: 5

Another study demonstrated the use of in situ nanomaterial synthesis in workwear fabrics [19]. Here, cotton-polyester twill fabric (70%–30% and 246 gsm) was coated in situ with nano-titania in in four steps; washing the fabrics, nano-titania preparation, injection of nano-titania into fabrics, and drying. The results of SEM, XRD, and DLS confirmed the in situ formation of nano-titania on the textile fibers. The UPF value of the uncoated and coated fabrics was 3.67 and 55.82, respectively. The in situ deposition of nano-titania on fabric can provide the required level of UV protection. Also, the intrinsic properties of the textile were not affected due to the nanofinishing process. Hence, it was concluded that the UV protective property of workwear fabrics can be improved by deposition of nano-titania.

3 In Situ Synthesis of Nanoparticles on Woollen Textiles

Montes-Hernandez et al. (2021) attempted in situ formation of Ag-NPs in wool. The wool fiber was first immersed in a silver ion solution (1 g/L of $AgNO_3$) for ion saturation at room Temperature for 24 h followed by draining fibers and their re-immersion this time in a strong chemical reducing solution (0.25 g/L of $NaBH_4$) for 24 h. This latter step leads to the in situ formation of Ag-NPs where size (5 nm < size <50 nm). This simple lab chemical method allowed instantaneous in situ formation of Ag-NPs onto fibers without the requirement of additional thermal treatment. In complement, all textile fibers containing Ag-NPs (sheep's wool 10 mg/g was submitted to interact with strong oxidants in an aqueous media (7.5% v/v of H_2O_2, 0.5 and 0.05 M of HNO_3 and ultrapure water as the control) using flow-through reactor experiments. In summary, this fundamental study indicated that Ag-NPs may be successfully adhered/inserted in wool in a safety-design perspective with performant biocide properties as confirmed by using *Bacillus subtilis*.

Montazer et al. [13] used in situ synthesis of nano-ZnO on wool. Zinc acetate (10%) was dissolved in 100 mL water at 30 °C under vigorous stirring. The wool fabric was immersed in the solution and remained till the end of the treatment. The temperature was then raised as the ammonia was being added under constant stirring, drop by drop into the solution (pH > 9) until the temperature reached 90 °C and then NaOH solution was added dropwise to the solution, with gentle stirring over a period of time to keep the pH above 9. The wool fabric was finally treated at 90 °C for 1 h. The presence of ZnO and $Zn(OH)_2$ possibly lead to interaction between protein chains of wool producing bonds between protein chains and increasing tensile strength. Nano-ZnO activated under daylight irradiation produced a permanent hydrophilic surface with lower yellowing and higher tensile strength.

4 In Situ Nanofinishing of Jute Textiles

Ammayappan et al. [2] treated the jute fabric in a bath containing 10 mM silver nitrate at pH 6.5. The material-to-liquor ratio was maintained at 1:30. The fabric was first kept at 40 °C for 15 min. After reaching 90 °C, the fabric was treated for 60 min with constant agitation to synthesize silver nanoparticles in situ. The silver nanoparticles remained stable on the jute fabric for up to 15 home washes. After that, 50% of them were stripped off due to the weakening of interactions between the jute fiber's functional groups and the nanoparticles. However, jute fiber naturally has the ability to synthesize silver nanoparticles in situ and retain them on its surface. The jute-silver fabric produced by this method exhibited good antibacterial properties and maintained semi-durability against home laundering.

In a reported research work (Joana et al., 2021), jute fabrics were functionalized with CaO and SiO_2 NPs and several functionalities were studied. In a first stage, the pre-treated fabrics were functionalized with CaO and SiO_2 NPs by a very simple in situ synthesis. For the in situ synthesis of only the CaO NPs onto the jute fabrics, 10 ml of a NaOH aqueous solution (1 M) were added drop by drop (1 ml/min) in 50 mL of a $Ca(NO_3)_2$ aqueous solution (0.2 M), under mechanical stirring at 50 °C for 1 h. Enhancement of the antimicrobial properties was observed with the addition of the CaO and SiO_2 NPs, reaching values of 99.8% of antibacterial activity against the tested microorganisms. The developed samples also showed photocatalytic activity, being able to degrade methylene blue (MB) dye. The presence of the SiO_2 NPs improved the obtained results as the MB degradation was faster for the sample with these NPs, when compared with the samples with only CaO NPs.

5 In Situ Nanofinishing of Linen Textiles

To synthesize cerium oxide NPs using in situ technique, Tripathi et al. [27] initially pre-treated the linen fabrics with chitosan followed by immersion in a 1% (w/v) cerium sulfate solution, for 45 min which serves as the nanoparticle precursor. After this, the fabric was removed and subjected to treatment with 0.1 N NaOH at 50 °C for 30 min under continuous sonication. The modified linen fabric was having multiple functionalities such as antibacterial activity UV protection, flame retardancy, and wrinkle resistance. The antimicrobial property is primarily due to the presence of chitosan and cerium cations (Ce^{3+}). Chitosan contains NH_2 groups that bind to negatively charged residues on the bacterial cell surface. This interaction disrupts the cell membrane, increasing permeability and causing leakage of intracellular substances, which inhibits bacterial growth. Similarly, Ce^{3+} followed a comparable mechanism to reduce microbial growth. The combined effect of chitosan and Ce^{3+}enhances the treatment's efficiency, resulting in a strong bactericidal effect.

In another study (Gupta et al., 2021), linen fabric was first treated with chitosan and then coated with ZnO nanoparticles. *Bombax ceiba* flower extract was used as both a reducing and capping agent for nanoparticle synthesis. This eco-friendly process resulted in a multifunctional fabric with enhanced textile properties. The chitosan modified linen fabric was treated with 0.5% zinc chloride solution for 1 h and then with the *Bombax ceiba* extract (5%) under ultra-sonication for 4 h. The in situ-synthesized ZnO nanoparticles improved the fabric's UPF to 26 before washing and 22 after 20 washes. The fabric also retained excellent antibacterial and antioxidant properties even after 20 wash cycles.

6 In Situ Nanofinishing of Silk Textiles

In a recent research work [10], silver nitrate as a precursor and amino acids from silk as reducing agents was used. Ag-NPs were synthesized in situ on the surface of silk without requiring additional reducing agents and catalysts. The antibacterial properties of the samples were evaluated against *Escherichia coli*, indicated excellent inhibition up to 99.91%.

7 Advantages of In Situ Nanofinishing

One of the key benefits of in situ nanofinishing is the strong interaction between nanoparticles and cotton fibers, preventing their leaching during washing or wear. Unlike conventional dip-coating or padding methods, which may lead to uneven nanoparticle distribution, the in situ approach ensures uniform coverage and deeper penetration into fiber structures. This results in enhanced durability, wash resistance, and long-term functionality of the treated fabric. Furthermore, in situ nanofinishing is an eco-friendly and cost-effective alternative to traditional textile finishing techniques. Since the nanoparticles are formed directly on the fabric, there is minimal waste generation, reducing the environmental footprint of the process. Additionally, the use of green reducing agents (e.g., plant extracts, natural polymers, and enzymes) in nanoparticle synthesis further enhances the sustainability of this approach.

In situ nanofinishing has found applications in a wide range of textile products, particularly in the medical, sportswear, and protective clothing industries. Antimicrobial cotton textiles treated with in situ-synthesized silver or copper nanoparticles are widely used in hospital linens, surgical gowns, and face masks to prevent microbial contamination. For UV protective clothing, in situ deposition of metal oxide nanoparticles such as ZnO and TiO_2 enhances the sun-blocking ability of cotton fabrics, making them suitable for outdoor and sportswear applications [3]. Similarly, hydrophobic and self-cleaning textiles can be developed using in situ nanofinishing with silica or fluorinated nanoparticles, which repel water and stains, keeping the fabric clean for longer durations. Another significant application is in the development of flame-retardant textiles, where in situ-synthesized nanoparticles such as magnesium hydroxide or phosphorus-based compounds improve the fire resistance of cotton fabrics. These textiles are particularly useful in workwear, military uniforms, and home furnishings to enhance safety against fire hazards.

Another study demonstrated the use of in situ coating of nano-Ag and nano-Cu on textiles for EMI shielding applications [1]. In this, before copper electroplating, the fabrics were activated with the in situ deposition of silver and copper particles. The lowest value of electrical resistivity was observed for copper plating over silver coated fabrics. The particles filled the micro spaces in between the fibers and made the continuous network to support the electrical conduction. It was reported that EMI shielding was increased with an increase in the deposition rate of the metal. Also, even after severe washing, there was an insignificant increase in electrical resistivity and good retention of the metal coating.

8 Challenges and Future Perspectives

Despite its numerous advantages, in situ nanofinishing still faces challenges related to scalability, cost, and process optimization. Large-scale industrial implementation requires precise control over nanoparticle size, distribution, and stability to maintain consistent performance across textile batches. Additionally, the potential release of nanoparticles into the environment during washing or disposal raises concerns regarding eco-toxicity and sustainability. To address these challenges, research is ongoing to develop bio-based and biodegradable nanoparticles for in situ nanofinishing, ensuring safer and greener textile processing. Integrating sustainability principles into nanofinishing presents both opportunities and challenges. The use of eco-friendly, biodegradable, and non-toxic nanomaterials is essential to minimize environmental and health risks, aligning with the goals of sustainable textile production. Energy-efficient processing techniques and the reduction of water and chemical usage are crucial for decreasing the ecological footprint of finishing operations. Adopting life cycle assessment (LCA) approaches helps evaluate the long-term sustainability of these technologies. Furthermore, aligning nanofinishing practices with the United Nations Sustainable Development Goals (SDGs)—particularly SDG 6 (Clean Water and Sanitation), SDG 9 (Industry, Innovation and Infrastructure), SDG 12 (Responsible Consumption and Production), and SDG 13 (Climate Action)—can drive innovation in responsible manufacturing. Future directions should focus on scalable, cost-effective, and circular nanotechnologies that uphold both environmental standards and market demands, ensuring a holistic transition to sustainable textile processing.

The future of in situ nanofinishing of textiles is highly promising, driven by advancements in nanotechnology, sustainable chemistry, and smart textile applications. With increasing demand for functional textiles, in situ nanofinishing offers a cost-effective, durable, and eco-friendly approach to impart properties such as antimicrobial activity, self-cleaning, UV protection, water repellency, and conductivity. Researchers are exploring green synthesis methods, utilizing plant-based or bio-inspired reducing agents to minimize environmental impact while ensuring controlled nanoparticle deposition on fibers. The development of stimuli-responsive nanocoatings that can change properties in response to temperature, pH, or external fields is another exciting avenue. Industries such as healthcare, military, fashion, and sportswear are adopting in situ nanofinishing to create antibacterial hospital linens, smart wearables, and stain-resistant fabrics. The integration of nanoparticles with phase-change materials can enhance thermal regulation, while conductive nanocoatings pave the way for e-textiles with embedded sensors and energy-harvesting capabilities. Regulatory advancements and standardization of nanoparticle safety protocols will play a crucial role in shaping commercialization. Future research will focus on improving scalability, wash durability, and biocompatibility, ensuring that in situ nanofinishing remains a key innovation in sustainable, high-performance textiles.

9 International Standards Governing nanofinishing of Textile Materials

Several international standards govern the nanofinishing of textiles, ensuring quality, safety, and performance. The International Organization for Standardization (ISO) has developed key standards such as ISO/TR 11360, which provides terminology and classification for nanomaterials, and ISO 31850, which outlines safety considerations for textiles incorporating nanomaterials. Additionally, ISO 10993 addresses the biocompatibility of nanocoated medical textiles. The American Society for Testing and Material (ASTM) has relevant standards like ASTM E2524, which evaluates the antimicrobial activity of silver nanoparticle-treated textiles, and ASTM E3025, which assesses the durability of nanoparticle coatings on fabrics. The OEKO-TEX® Standard 100 and (Registration, Evaluation, Authorisation, and Restriction of Chemical (REACH) regulations in the EU ensure that nanofinished textiles meet environmental and health safety criteria. These standards collectively regulate aspects like nanoparticle release, durability, toxicity, and functional performance to ensure the safe and effective use of nanotechnology in cotton textiles.

10 Conclusion

The integration of in situ nanofinishing in textiles represents a transformative advancement in textile engineering, redefining the functional and aesthetic attributes of natural fiber fabrics. This cutting-edge approach enables the direct synthesis of nanoparticles on fiber surfaces, ensuring enhanced durability, superior performance, and multifunctionality without compromising the inherent qualities of cotton. This makes it an ideal approach for developing high-performance textiles that cater to the evolving demands of consumers and industries alike. Despite its immense potential, the adoption of in situ nanofinishing techniques faces certain challenges, including scalability, cost-effectiveness, and environmental concerns associated with nanoparticle synthesis. However, continuous advancements in green chemistry, sustainable process engineering, and nanotechnology are addressing these limitations by developing eco-friendly, energy-efficient, and economically viable solutions. Innovations such as the use of bio-based nanoparticles, water-based synthesis methods, and resource-efficient fabrication techniques are paving the way for a more sustainable and responsible approach to textile finishing. As the textile industry moves toward a more sustainable future, the convergence of nanotechnology with natural fiber textiles through in situ nanofinishing will play a pivotal role in shaping the next generation of high-performance fabrics. The synergy between scientific research, technological innovation, and industrial application will not only enhance the functionality and longevity of textiles but also contribute to reducing the environmental footprint of textile processing. By embracing this sustainable and cost-effective approach, the industry can unlock new opportunities for the development of advanced textile materials that meet both performance requirements and sustainability goals. In conclusion, in situ nanofinishing of cotton textiles is not merely an enhancement technique but a paradigm shift in textile manufacturing. Its ability to seamlessly integrate multifunctionality with sustainability positions it as a key enabler of next-generation textile solutions. With continued research and technological advancements, in situ nanofinishing holds the promise of revolutionizing textile finishing, driving the industry toward more eco-friendly, durable, and high-performance fabric innovations.

References

1. Ali, A., Hussain, F., Kalsoom, A., Riaz, T., Zaman Khan, M., Zubair, Z., Shaker, K., Militky, J., Noman, M. T., & Ashraf, M. (2021). Multifunctional electrically conductive copper electroplated fabrics sensitizes by in-situ deposition of copper and silver nanoparticles. *Nanomaterials, 11*, 3097.
2. Ammayappan, L., & Sujay, C. (2017). Coating of silver nanoparticles on jute fibre by in situ synthesis. *Cellulose, 24*, 1563–1577. https://doi.org/10.1007/s10570-017-1204-2
3. Arputharaj, A., Nadanathangam, V., & Shukla, S. R. (2020). Development of multi-functional cotton surface for sportswear using nano zinc oxide. *Journal of Natural Fibers, 17*(3), 346–358.
4. Arputharaj, A., Nadanathangam, V., & Shukla, S. R. (2017). A simple and efficient protocol to develop durable multifunctional property to cellulosic materials using in situ generated nano-ZnO. *Cellulose, 24*, 3399–3410.
5. Soleimani-Gorgani, A., Avinc, O., & Alborz, R. (2023). Sustainable antibacterial cotton fabrics with in situ formed silver nanoparticles by bio-inkjet printing. *Journal of Cleaner Production, 386*, 135796.
6. Bayisa, T., Deressa, G., Feyisa, Z., Guta Inki, L., & Gupta, N. K. (2024). In-situ synthesis of CuO/TiO2 nanocomposite onto amine modified cotton fabric for antibacterial durability and UV protection. *Journal of Natural Fibers, 21*(1), 2346120.

7. Chen, W., Feng, X., Zhang, D., Lu, F., Wang, H., Tan, J., Xu, Q., Liu, Y., Cao, Z., & Su, X. (2022). In situ synthesis of TiO_2/NC on cotton fibers with antibacterial properties and recyclable photocatalytic degradation of dyes. *RSC Advances, 12*, 19974–19980.
8. Javed, A., Wiener, J., Tamulevičienė, A., Tamulevičius, T., Lazauskas, A., Saskova, J., & Račkauskas, S. (2021). One step in-situ synthesis of zinc oxide nanoparticles for multifunctional cotton fabrics. *Materials, 14*, 3956.
9. Jazbec, K., Šala, M., Mozetič, M., Vesel, A., & Gorjanc, M. (2015). Functionalization of cellulose fibres with oxygen plasma and ZnO NPs for achieving UV protective properties. *Journal of Nanomaterials, 16*, 25.
10. Jin, X. Q., Li, J. L., Liu, J., et al. (2025). In situ synthesis of silver nanoparticles on silk: Producing antibacterial fabrics. *Bioprocess and Biosystems Engineering*. https://doi.org/10.1007/s00449-025-03138-0
11. Jur, J., Sweet, W. J., Oldham, C. J., & Parsons, G. N. (2011). Atomic layer deposition of conductive coatings on cotton, paper, and synthetic fibres: Conductivity analysis and functional chemical sensing using "all-fibre" capacitors. *Advanced Functional Materials, 21*, 1993–2002.
12. Mao, Z., Shi, Q., Zhang, L., & Cao, H. (2009). The formation and UV-blocking property of needle-shaped ZnO nanorod on cotton fabric. *Thin Solid Films, 517*, 2681–2686.
13. Montazer, M., Amiri, M. M., & Malek, R. A. M. (2013). In situ synthesis and characterization of nano ZnO on wool: Influence of nano photo reactor on wool properties. *Photochemistry and Photobiology, 89*(5), 1057–1063.
14. Montazer, M., Alimohammadi, F., Shamei, A., & Rahimi, M. K. (2012). In situ synthesis of nano silver on cotton using Tollens' reagent. *Carbohydrate Polymers, 87*, 1706–1712.
15. Montes-Hernandez, G., Di Girolamo, M., Sarret, G., Bureau, S., Fernandez-Martinez, A., Lelong, C., et al. (2021). In situ formation of silver nanoparticles (Ag-NPs) onto textile fibers. *ACS Omega, 6*(2), 1316–1327.
16. Nakamura, S., Sato, M., Sato, Y., Ando, N., Takayama, T., Fujita, M., & Ishihara, M. (2019). Synthesis and application of silver nanoparticles (Ag NPs) for the prevention of infection in healthcare workers. International journal of molecular sciences, 20(15), 3620.
17. Perelshtein, I., Applerot, G., Perkas, N., Wehrschetz-Sigl, E., Hasmann, A., Guebitz, G. M., et al. (2008). Antibacterial properties of an in situ generated and simultaneously deposited nanocrystalline ZnO on fabrics. *ACS Applied Materials and Interfaces, 1*, 361–366.
18. Prasad, V., Arputharaj, A., Bharimalla, A. K., Patil, P. G., & Vigneshwaran, N. (2016). Durable multifunctional finishing of cotton fabrics by in situ synthesis of nano-ZnO. *Applied Surface Science, 390*, 936–940.
19. Rabiei, H., Farhang-Dehghan, S., Montazer, M., Khaloo, S. S., & Koozekonan, A. G. (2022). UV protection properties of workwear fabrics coated with TiO2 nanoparticles. *Frontiers in Public Health, 10*, 929095.
20. Román, L. E., Amézquita, M. J., Uribe, C. L., Maurtua, D. J., Costa, S. A., Costa, S. M., Keiski, R., Solís, J. L., & Gómez, M. M. (2020). *Advances in Natural Sciences: Nanoscience and Nanotechnology, 11*, 025009.
21. Sadr, F. A., & Montazer, M. (2014). In situ sonosynthesis of nano TiO_2 on cotton fabric. *Ultrasonics Sonochemistry, 21*(2), 681–691.
22. Salama, M., & El-Sayed, A. A. (2014). Imparting permanent antibacterial properties to viscose fabric using zinc oxide nanoparticles and polymeric binders. *World Applied Sciences Journal, 32*, 392–398.
23. Shateri-Khalilabad, M., & Yazdanshenas, M. E. (2013). Bifunctionalization of cotton textiles by ZnO nanostructures: Antimicrobial activity and ultraviolet protection. *Textile Research Journal, 83*, 993–1004.
24. Shishoo, R. (2007). Introduction—Potential application of plasma technology in the textile industry. In R. Shishoo (Ed.), *Plasma technologies for textile*. Woodhead Publishing.
25. Tang, B., Lin, X., Zou, F., Fan, Y., Li, D., Zhou, J., Chen, W., & Wang, X. (2017). In situ synthesis of gold nanoparticles on cotton fabric for multifunctional applications. *Cellulose, 24*, 4547–4560.

26. Tania, I. S., Ali, M., & Azam, M. S. (2019). In-situ synthesis and characterization of silver nanoparticle decorated cotton knitted fabric for antibacterial activity and improved dyeing performance. *SN Applied Sciences, 1*, 64.
27. Tripathi, R., Narayan, A., Bramhecha, I., & Sheikh, J. (2019). Development of multifunctional linen fabric using chitosan film as a template for immobilization of in-situ generated CeO_2 nanoparticles. *International Journal of Biological Macromolecules, 121*, 1154–1159.
28. Uğur, Ş. S., Sarıışık, M., Aktaş, A. H., Uçar, M. Ç., & Erden, E. (2010). Modifying of cotton fabric surface with nano-ZnO multilayer films by layer-by-layer deposition method. *Nanoscale Research Letters, 5*, 1204.
29. Verbič, A., Šala, M., Jerman, I., & Gorjanc, M. (2021). Novel green in situ synthesis of ZnO nanoparticles on cotton using pomegranate peel extract. *Materials, 14*, 4472.
30. Vigneshwaran, N., Kumar, S., Kathe, A. A., Varadarajan, P. V., & Prasad, V. (2006). Functional finishing of cotton fabrics using zinc oxide-soluble starch nanocomposites. *Nanotechnology, 17*, 5087–5095.
31. Vitchuli, N., Shi, Q., Nowak, J., Kay, K., Caldwell, J. M., Breidt, F., et al. (2011). Multifunctional ZnO/Nylon 6 nanofibre mats by an electrospinning–electrospraying hybrid process for use in protective applications. *Science and Technology of Advanced Materials, 12*, 055004.
32. Zhou, Q., Chen, J., Lu, Z., Tian, Q., & Shao, J. (2022). In situ synthesis of silver nanoparticles on flame-retardant cotton textiles treated with biological phytic acid and antibacterial activity. *Materials, 15*, 2537.

Sustainable Nanotextiles: A Review on Innovations and Challenges

Ramachandran R. and Madireddy Sandeepa

Abstract Nanotechnology is an emerging field covering a range of scientific and technological categories, moving forward in the textile industry. For instance, the ability to manipulate materials at nanoscale one billionth of a meter has led to the fabrication of textiles with desirable features like stain resistance, wrinkle-free properties, static-eliminating properties, and electrical conductivity while maintaining comfort and flexibility. This article explores the impact of nanotechnology on textiles, especially sustainable innovations. Nanomaterials can also be integrated into fabrics to develop multifunctional fabrics, such as antimicrobial, UV resistant, flame-retardant, and electrically conductive, leading to more durable and sustainable products. In addition, smart nanomaterial-based devices enable textiles to react to external factors such as a change in electrical signal, color or physiological state, contributing to the development of connected clothes with potential applications across different sectors such as fashion, healthcare, defense, and sports. The review notes the growing demand for high-performance sustainable textiles and gives insights on the approaches to integrate nanomaterials affordably and sustainably. It also discusses the potential health and environmental risks associated with nanotextiles, including nanotoxicity and their ecological consequences as well as the importance of conducting life cycle assessments (LCAs) of new nanotechnology and developing green nanotechnology. The current problems in developing green and biocompatible nanomaterials along other potential applications of smart textiles in energy harvesting, environmental monitoring, and sustainability are also highlighted. In a timely context, this chapter discusses current trends and challenges focusing on functionalizing textiles with nanomaterials to be in line with societal expectations of safety, cost-effectiveness, and sustainability alongside addressing potential health and environmental issues. The future of smart textiles utilizing nanomaterials for energy conversion and environmental monitoring is also discussed alongside effective waste disposal practices, emphasizing green nanotechnology's ability to avoid harmful effects. Finally, it highlights the challenges of

Ramachandran R. (✉) · M. Sandeepa
National Institute of Fashion Technology, Kannur, Kerala, India
e-mail: ramachandran.r@nift.ac.in

S. S. Muthu (ed.), *Sustainable Nano Textile Technology*, Textile Science and Clothing Technology, https://doi.org/10.1007/978-3-032-05902-4_3

integrating such green nanomaterials in these innovations and the need for further research toward greener environment emphasizing the importance of advancing textiles safely with limited adverse effects.

Keywords Nanotechnology · Sustainability · Nanomaterials · Smart nanotextiles

1 Introduction

Nanotechnology involves the manipulation of matter at the extremely small scale, specifically within the 1–100 nm range, which is one-billionth of a meter. This scale enables the observation and engineering of materials with unique properties different from their bulk counterparts. For context, a human nail is approximately a billion nanometers (1 m = 1,000,000,000,000 nm), and the diameter of a human hair varies between approximately 75,000 and 100,000 nm, illustrating the vast difference in sizes at the nanoscale. The conceptual foundation of nanotechnology was laid toward the end of the twentieth century, with significant advances occurring around the 1970s. The U.S. National Nanotechnology Initiative (NNI) defines nanotechnology as research and development at the atomic, molecular, or macromolecular scale within the 1–100 nm range. At this scale, materials exhibit novel electrical, optical, chemical, and biological properties, fostering innovation across multiple industries. The integration of nanomaterials into fabrics is revolutionizing textile performance by providing properties such as enhanced durability, self-cleaning, antimicrobial effects, UV protection, and sensory capabilities. These advancements support the development of high-performance, sustainable textiles, and smart wearable systems. This exploration underscores how nanotechnology-driven innovations are shaping the future of textiles, addressing challenges such as environmental sustainability, safety, and scalability, while opening opportunities in energy harvesting, environmental monitoring, and the creation of intelligent textiles.

1.1 Nanotechnology and Daily Life Textiles

Nanoparticles are ubiquitous in the environment, with measurements indicating approximately 100,000 nanoparticles per cubic centimeter (cm^3) of air in urban centers and about 50,000 nanoparticles per cm^3 in forested areas. Figure 1 depicts a wide range of uses of nanotechnology in various industries, including how it can improve material qualities, advance energy and defense technologies, and improve medical and agricultural solutions. The design graphically highlights nanotechnology's multidisciplinary character and its increasing significance in contemporary innovation. This widespread presence underscores the natural occurrence of nanoscale particles and their increasing relevance in diverse fields, including textiles. The pervasive nature of nanoparticles demonstrates both their acceptability

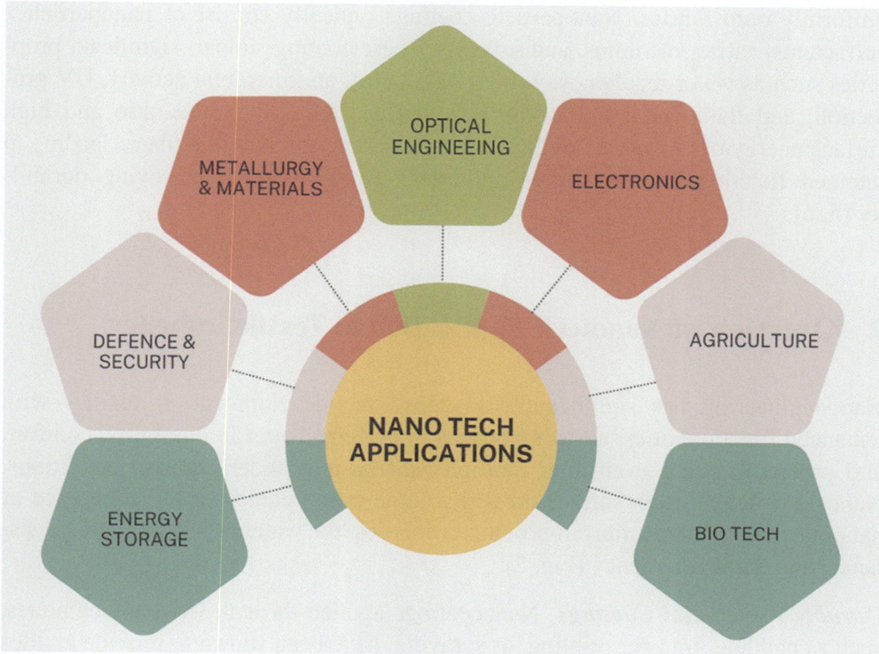

Fig. 1 Application of Nanotechnology

and potential for integrated applications across various sectors [4]. Traditional natural fibers like cotton are valued for their absorbency, softness, and breathability. However, they tend to lack strength, durability, and are prone to wrinkling and staining. Synthetic fibers can be engineered to be antimicrobial or stain-resistant but often compromise comfort compared to natural fibers [4]. Nanotechnology enables the enhancement of textile properties without sacrificing fundamental qualities such as breathability and tactile feel ("hand"). By integrating nanomaterials into fibers, manufacturers can create multifunctional textiles that meet the rising consumer demand for improved appearance, color retention, shape stability, and additional functionalities [2, 3].

1.2 Evolution of Nanotechnology in Textiles

The pioneering efforts in nanotechnology for textiles were initiated by Nano-Tex, a subsidiary of Burlington Industries based in the USA. Nano-Tex introduced nanotechnology-based treatments to enhance fabric functionalities, setting the stage for subsequent industry investments and innovations in the field [5]. Various techniques have been employed to apply nanoparticles onto textiles, including spraying, transfer printing, washing, rinsing, and padding. Among these, padding remains the most prevalent due to its effectiveness and ease of applying nanoparticle coatings uniformly onto fabrics. Nanoparticle coatings typically consist of nanoparticles, surfactants, carrier mediums, and solvents. These coatings impart significant properties such as water repellency, wrinkle resistance, antimicrobial activity, UV protection, and flame retardancy. The high surface area-to-volume ratio and high surface energy of nanoparticles enhance their adherence to textile fibers, leading to retained functionality even after multiple washes, thereby improving durability [5, 6].

2 Overview of Nanotech Application in Textile Industry

Nanoengineering has significantly advanced the development of textiles with enhanced functionalities, allowing fabrics to perform better in durability, comfort, and protective roles, even after multiple washes and under abrasive conditions. Nanotechnology's high surface area-to-volume ratio enables strong adherence of nanomaterials to fibers, thus improving properties like water repellency, stain resistance, and flame retardancy (Fig. 2).

Durable Functional Coatings Nanocoatings applied through various techniques, such as padding, transfer printing, or spraying, have been shown to provide textiles with water resistance, stain repellency, and fire retardancy. These coatings extend

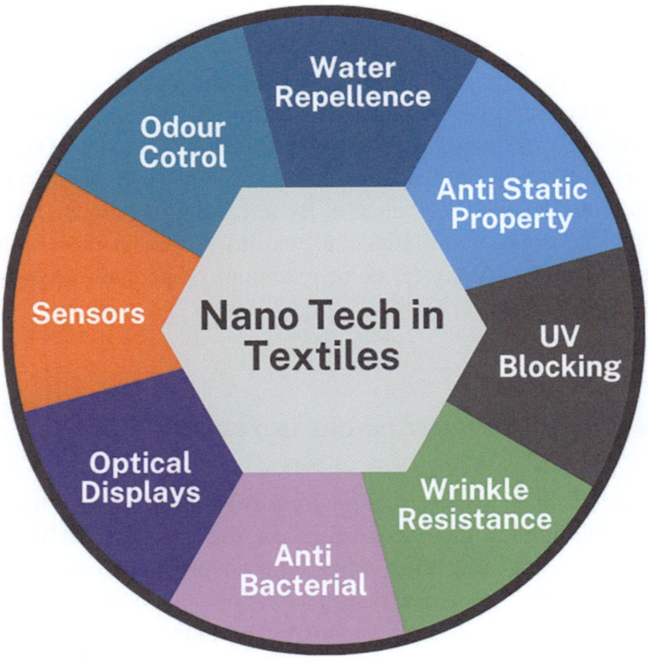

Fig. 2 Nanotechnology in textile

the longevity and performance of textiles by maintaining their functional properties over time despite laundering and wear.

Antimicrobial Properties Silver nanoparticles are notably effective antimicrobial agents, which impart odor control and reduce microbial growth on fabrics, thus extending their usability and maintaining hygiene [8, 10].

Conductivity and Wearable Electronics The integration of conductive nanomaterials like carbon nanotubes or graphene forms the foundation for electronic textiles ("e-textiles") and wearable devices capable of monitoring physiological parameters, offering applications in healthcare, sports, and defense sectors [9].

Aesthetic and Color-Functional Nanotechnology Nanotechnology enables the development of color-tunable fabrics and new dyeing techniques, expanding creative opportunities for designers without compromising fabric quality or environmental safety.

UV Protection Nanoparticles such as titanium dioxide (TiO_2) provide efficient UV-protective properties, shielding wearers from harmful ultraviolet rays while maintaining fabric transparency and comfort.

Thermal Regulation Nanoengineered textiles can interact with environmental heat transfer, achieving thermal regulation by either insulating or dissipating heat as needed, ensuring wearer comfort across different conditions [7].

Smart Textiles and Wearable Devices Garments embedded with flexible electronic components and sensors can monitor health metrics or environmental conditions in real time, facilitating applications in medical health monitoring, sports performance, and defense operations [8].

2.1 Nanotechnology and Fashion

Nanotechnology also comes into play in providing wearable technology and intelligent clothing in fashion. The designers and innovators have made good use of nanotechnology in clothing designs that do not compromise style for functionality. Examples include dresses created by Philips that change color depending on the wearer's mood and performance apparel with organic light-emitting diodes (OLEDs) of artists like the Black-Eyed Peas. By example, the iconic designs of Studio XO and Tech Haus made a big splash in allowing nanotechnology to bring forth expression through clothing that reacts to stimuli, such as Lady Gaga's 3D-printed bubble machine dress and drone dress. Wearable technology has revolutionized sports apparel, healthcare, and professional uniforms. Nanoengineered textiles would monitor physiological parameters in real time to enhance performance and safety.

For example, nano-coated fireproof uniforms offer protection for firefighters without impacting their comfort and weight. Nanotechnology also adds to the quest for sustainability in fashion. It will make the clothes more durable and effectively self-cleaning, thus reducing the frequency of washes and prolonging the life of garments, in line with the increasing push toward sustainability in the fashion sector.

3 Innovations in Nanotextiles

3.1 Stain Resistance

Nanotechnology has progressed the development of stain-resistant fabrics by modifying the textile surfaces at a molecular level, which repels liquids and prevents stains. Among the most famous applications is nanowhisker incorporation into fabrics, creating a superhydrophobic effect in which liquids tend to bead up and roll off the material. This technology preserves the breathability and comfort of the fabric but provides effective stain resistance. Besides, plasma treatments have been applied to deposit hydrophobic nanofilms on fabric surfaces for the enhancement of stain resistance. These treatments modify the surface energy of textiles in such a way that it repels water and other liquids. Nanotechnology application in textiles improves the stain resistance properties of textiles and contributes to other functional effects, such as antimicrobial activity, UV protection, and durability. Thus, it plays a versatile role in modern textile engineering. Commercial companies like Nanotex have perfected such technologies with fabrics that can be water- and stain-resistant without losing the drape or hand. It applies its treatment to a lot of products—from apparel and home goods up to increasing the durability and the ease of their care.

3.2 Wrinkle Resistance

The conventional wrinkle-resistant treatments tend to deteriorate the inherent properties of fabrics, including tensile strength and breathability. Nanotechnology presents an alternative avenue for such treatment using materials such as nano-TiO_2 and nano-silica. Nano-TiO_2 catalyzes cross-linking reactions under UV light to improve the wrinkle resistance of cotton. Nano-silica, combined with maleic anhydride, enhances the wrinkle resistance of silk without deteriorating its quality.

3.3 Anti-Bacterial Property

Nano-sized silver, titanium dioxide, and zinc oxide particles provide antibacterial activity in textiles. The high surface area of these particles maximizes sterilization effects. Nanosilver interacts with proteins in bacteria and fungi, which can disrupt cellular processes and inhibit growth. It is widely used in socks and medical products such as wound dressings. Titanium dioxide is a photocatalyst that, upon UV light irradiation, forms reactive oxygen species that break down organic matter, including bacteria. ZnO works similarly as a photocatalyst with a slightly different band gap but shows similar antibacterial properties. The treated fabrics are more resistant to microbial growth and stain discoloration.

3.4 UV Resistant

TiO_2 and ZnO, among other inorganic UV blockers, are more commonly used instead of organic alternatives as they are nontoxic and have stability toward chemical changes during UV and high-temperature exposures. Nano-sized TiO_2 and ZnO perform better compared to the normal material for UV absorption and scattering, which are because of its large surface area. Rayleigh scattering theory assumes that a size range between 20 and 40 nm is perfect for the UV scattering effect.

Nanotechnology-based UV-blocking treatments, such as the sol-gel method, coat fabrics with a thin layer of TiO_2, providing excellent UV protection that endures up to 50 washes. Similarly, ZnO nanorods, ranging from 10 to 50 nm in length, have demonstrated superior UV protection, achieving high ultraviolet protection factor (UPF) ratings when applied to cotton fabrics.

3.5 Water Repellent

Nanotechnology has revolutionized the hydro-phobic character of fabrics. For example, Nano-Tex uses nanowhiskers which are hydrocarbon structures invented at 1/1000 the size of a single cotton yarn. These whiskers work in a way that produces a "peach fuzz" effect, enabling water repulsiveness but not less the strength of cotton. The nanowhiskers spacing is smaller than water droplet and greater than water molecules, so water will rest on the nanowhiskers unless pressure is applied and then it washes off. The application is permanent and has allowed the textile to retain its porosity. As well, the Swiss fabrics firm Schoeller developed NanoSphere technology as an improvement in water repellency. The concept emulates the lotus effect through three dimensional surface and gel-forming additives which aids in water repel and dirt adhesion prevention. The superhydrophobic characteristics of lotus leaf surfaces turn water droplets into spherical shapes, which roll off and take dirt along with them, and when it rains, the surface remains dry. In addition, plasma nanotech could also make cotton fabric hydrophobic in such a way as to apply a nanoparticulate fluorocarbon plasma film on the cotton fabric. This strategy has increased the water repellency characteristic with no alteration of fabric softness that is usually noted to accompany the water resistant fiber.

3.6 Flame Retardent

Nanotechnology has significantly advanced flame retardant technologies by enabling the development of materials that enhance fire resistance through molecular-level modifications. These innovations have led to more effective and environmental-friendly flame retardant solutions across various industries. *Carbon nanotubes (CNTs)*: Incorporation of carbon nanotubes into polymers has been demonstrated to enhance fire resistance. When CNTs are exposed to heat, they create a continuous layer of carbon that coats the surface of the material, thereby protecting it as a thermal shield to slow down degradation and improve its capability to withstand elevated temperatures.

Graphene-Based Nanomaterials (GNMs): Graphene and its derivatives have been considered as flame retardants owing to its superior thermal properties. GNMs can create a protective shield on the surface of materials, preventing flame spread and thereby reducing heat release. Studies suggest that the incorporation of GNMs in polymers improves radically their flame retardancy. *Nano-Flame Retardants for Unsaturated Polyester Resins (UPRs)*: Studies show that nanoscale flame retardants applied at low loadings onto UPRs can significantly improve thermal stability with minimal impact on mechanical properties. They are hence suitable for applications requiring both strength and fire resistance. The incorporation of nanoclays into polymers has been reported to enhance flame retardancy. Nanoclays can create a char layer on the surface of the material during combustion, which acts as a barrier to slow heat release and gas diffusion. This mechanism enhances the resistance of the material to ignition and flame spread. Companies have developed fireproof coatings using nanotechnology that offer superior flame spread resistance and significantly reduced heat transfer. These coatings can be applied to various substrates, offering enhanced fire protection for building materials and other applications. These developments in nanotechnology-based flame retardants open promising avenues for developing safer, more efficient, and environmentally friendly fire-resistant materials.

3.7 Smart NanoTextiles

A "smart" system is often defined as a system engineered to sense, control, and respond to external stimuli in its environment. Smart textiles are defined as fabrics that are able to sense and interact with external conditions or stimuli, hence giving an appropriate response. The integration of fabrics with advanced nanomaterials is the future generation of intelligent textiles. Figure 3 shows the smart nanoapplications as textile-integrated fabrics. These materials make it possible for textiles to sense, protect against hazards, connect to devices, and adapt to environmental changes. The earliest prototype of intelligent textiles was the mooring jacket. It was wearable electronic apparel with ICD-integrated outerwear.

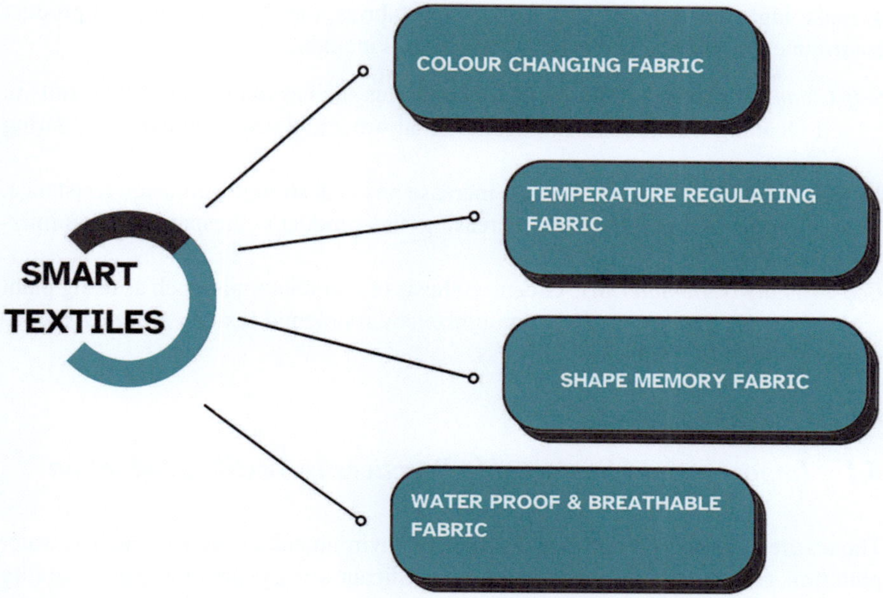

Fig. 3 Smart nanoapplication

Developed in the last century, it used traditional materials to carry integrated devices such as microphones, headphones, cellular phones, and MP3 devices. The manipulation of fabrics with smart nanoparticles has advanced the intelligent textile functionality. Smart textiles today are applied beyond clothing to military, defense, and healthcare sectors. For example, smart textiles are currently used in drug delivery systems for the treatment of skin diseases. Other inventions include temperature-controlled clothing, wearable energy harvesting based on the kinetic energy of human movement, solar-powered jackets, and e-textiles integrated with graphene to detect poisonous gases. Wearable e-textiles are one of the largest categories of intelligent textiles, which work as sensors and often have displays with light that will alert the user in situations involving changes in heartbeat, temperature, and blood pressure. These textiles are likely to transmit warning signals via mobile applications or wireless devices, thereby ensuring health monitoring and safety in emergency conditions.

4 Sustainability in NanoTextiles

Nanotechnology in textiles is changing the face of the industry, helping make the fabric production sustainable by changing the properties of fabrics and saving the resources. Nanotextiles are of nanomaterials and nanotechnology-based processes to make high-functionality and long-lasting fabrics, thereby reducing end product environmental impact. The main advancements include:

Self-Cleaning Fabrics: Addition of nanoparticles such as titania and ZnO results in self-cleaning conditions without additional water and detergent exposure during washing.
Increased Longevity: Nanomaterials increase physical strength, abrasion resistance, and chemical stability, thus increasing the product's lifespan and minimizing waste.
Eco-Friendly Nanomaterials: Green synthesis of nanomaterials, such as using plant extracts or enzymes, is used to minimize environmental toxicity and energy consumption in the production process.

4.1 Importance of Sustainable Practices in Textile Production

The textile industry is a significant source of environmental problems such as water pollution, greenhouse gas emissions, and significant waste generation. Sustainability must be adopted in order to counter these challenges.

Key elements include:

Resource Efficiency: Reduce the consumption of water, energy, and chemicals in the production process, especially in dyeing and finishing.
Circular Economy: Encourage recycling, upcycling, and biodegradability in order to minimize textile waste.
Biodegradable Materials: Using organic fibers such as cotton and hemp or recycled plastics and agricultural wastes.
Nontoxic Chemicals: Encouraging the application of low impact dyes, and avoiding substances that are considered harmful, like heavy metals.

Sustainable practices not only reduce environmental damage but also reflect consumer preference for eco-friendly products. They enable companies to get a competitive edge and ensure sustainable long-term success in the industry.

4.2 Use of Nanomaterials in Sustainable Innovations

Integrating nanotechnology into textiles enables innovations that align with sustainability goals, including improved performance with reduced resource usage. The key applications include silver nanoparticles and chitosan coatings provide antimicrobial properties, reducing odors and the need for frequent laundering. Nanomaterials like graphene facilitate energy storage and harvesting, using renewable sources like body motion or solar energy to power electronic devices embedded in textiles. The nanocoating improves resistance against UV, and the clothes become more protective of the human skin from ultraviolet rays and further extend the service life of clothing.

4.3 Health and Environmental Risks

Nanotextiles are the game changer of the textile world. Nano is changing the old fabric functionalities, and their development brings about numerous concerns over potential health and environmental issues. Due to their tiny dimensions and properties, nanoparticles are sometime dangerous to be used safely in products. Workers handling or using nanotextiles may be exposed to nanoparticle through inhalation or direct contact with skin. This might cause problems, such as respiratory discomfort or irritation in the skin. Nanoparticles, such as silver or titanium dioxide, might accumulate in tissues of humans with time and hence interfere with cell functions. Being a new textile, nano has no long-term health effects, so one has to be very careful about them. Environmental Concerns related to the usage of nanoparticles are also increasing. Nanotextiles might damage aquatic life or disrupt ecological systems when released into water from washing. Nanoparticles are so mobile and don't break down that easily, meaning they can persist and bioaccumulate in soil and water, affecting biodiversity and potentially entering food chains [9].

4.4 Nanotoxicity and Ecological Consequences

The nanotextiles have given rise to a new issue of nanotoxicity; the harmful effects' nanoparticles can have on living organisms and the environment. Their small size and unique properties make them unpredictable and, in some cases, harmful. Nanoparticles such as silver or zinc oxide can penetrate water bodies and cause harm to aquatic organisms' health and reproduction. Once nanoparticles penetrate the soil, they can disturb the balance of the microbial communities, which are important for nutrient cycling and general soil health. Further nanoparticles absorbed by smaller organisms can travel up the food chain, possibly affecting larger species, including humans.

Nanoparticles differ in their behavior as compared to any bulk material, and their toxicity may not be assessed predictively by the conventional methods. Standard protocols for testing are unavailable. And thus, it is tough to ascertain the long-term impacts and hazards which may arise to the environment.

4.5 Importance Of Life Cycle AssessmentS (LCAs) In Evaluating Nanotextiles

In order to get an authentic view about nanotextiles, one has to follow their whole lifecycle, which means from extraction of raw material till its end-of-life scenario. The following aspects are to be considered at every stage of life cycle.

Stages of LCA

Raw Materials: What amount of energy and resources goes into developing nanomaterials?
Production: What emissions, waste, and energy are associated with producing nanotextiles?
Usage: Impact of nanotextiles on the users and the surroundings throughout their life.
End-of-life: How are nanotextiles recycled or degraded? In this final phase, how are nanoparticles released in the environment?

The major advantages of LCA is it enables one to know where nanotextiles are most detrimental to the environment, and it supports sustainability through tacing eco-friendly production methods and moreover it balances their benefits with the possible health and environmental hazards. At the same time the challenges associated with LCA are available of limited data on the behavior of the nanoparticles at any lifecycle stages which makes it very difficult to evaluate their entire impact, and also Modelling such traces of nanoparticles can also prove to be tough and requires more research work.

Adoption of LCA by industries helps in developing a balanced approach that ensures nanotextiles are innovative and simultaneously environmentally responsible.

5 Green Nanotechnology

Green Nanotechnology refers to the design, production, and application of nanomaterials in a way that minimizes environmental impact and promotes sustainability [1]. It emphasizes the use of eco-friendly processes and materials to create nanotechnology solutions that are safe for human health and the environment.

Principles of Green Nanotechnology include:

Sustainable Synthesis: Utilizing renewable resources and environmentally benign methods for the synthesis of nanomaterials, such as biological systems (plants, bacteria, fungi) to reduce the use of toxic chemicals and energy consumption.

Minimization of Waste: Implementing processes that generate minimal waste and by-products during the production of nanomaterials, thereby reducing the ecological footprint.

Safety and Toxicity Assessment: Evaluating the potential health and environmental impacts of nanomaterials throughout their lifecycle, ensuring that they are safe for use and do not pose risks to ecosystems or human health.

Biocompatibility: Developing nanomaterials that are compatible with biological systems, making them suitable for applications in medicine and healthcare without adverse effects.

Energy Efficiency: Focusing on energy-efficient processes in the production and application of nanomaterials to reduce overall energy consumption and greenhouse gas emissions.

5.1 Strategies for Developing Biocompatible and Eco-Friendly Nanomaterials

Green Synthesis Approaches: Employing biological methods for the synthesis of nanomaterials, such as using plant extracts, algae, or microorganisms, which can act as reducing and stabilizing agents. This approach often requires milder conditions and results in less toxic by-products.

Use of Natural Polymers: Incorporating biopolymers (e.g., chitosan, cellulose, and alginate) in the formulation of nanomaterials to enhance biodegradability and biocompatibility, making them suitable for environmental and biomedical applications.

Functionalization with Biocompatible Agents: Modifying nanomaterials with biocompatible substances to improve their interaction with biological systems, enhancing their safety and effectiveness in medical applications such as drug delivery and tissue engineering.

Life Cycle Analysis (LCA): Conducting comprehensive assessments of the environmental impacts of nanomaterials from production to disposal, ensuring that they contribute positively to sustainability goals and do not harm the environment.

Recyclability and Reusability: Designing nanomaterials that can be easily recycled or reused, thus promoting a circular economy and reducing waste.

Regulatory Compliance and Standards: Adhering to environmental regulations and safety standards during the development and application of nanomaterials to ensure they meet safety and sustainability criteria.

By implementing these strategies, researchers and manufacturers can develop nanomaterials that are not only effective in their applications but also align with the principles of sustainability and environmental responsibility [4].

6 Current Trends and Challenges

The following tables give an overview of current trends and application of nanotechnology in textile and fashion domain.

Area	Trend	Applications
Antimicrobial and antiviral textiles	Incorporating nanoparticles like silver, zinc oxide (ZnO), and copper to provide antimicrobial and antiviral properties.	Medical textiles, sportswear, and personal protective equipment (PPE). Example: Silver nanoparticles embedded in masks to combat pathogens like COVID-19.
Self-cleaning fabrics	Using photocatalytic nanoparticles like titanium dioxide (TiO2) and hydrophobic nanocoatings to enable self-cleaning properties	Curtains, carpets, and outdoor gear.
Energy-harvesting textiles	Embedding nanomaterials like graphene and carbon nanotubes for piezoelectric or triboelectric energy harvesting.	Wearable electronics and smart textiles. Example: Fabrics that generate electricity from body movements.
UV protection	Incorporating UV-blocking nanoparticles like ZnO and cerium oxide (CeO2) to enhance UV protection	Outdoor apparel and sun-protective clothing.
Flame-retardant textiles	Using nanoscale additives like montmorillonite clay and carbon nanofibers for improved thermal resistance.	Firefighter uniforms and upholstery.
Conductive textiles	Embedding conductive nanomaterials like silver nanowires, graphene, and carbon nanotubes.	Flexible circuits, e-textiles, and wearable sensors. Example: Smart clothing for fitness tracking or medical diagnostics.
Smart and responsive textiles	Integrating nanotechnology for thermoregulation, color-changing, and moisture-wicking capabilities.	Fashion, sports, and healthcare. Example: Thermochromic fabrics that change color with temperature variations.
Sustainability and eco-friendly textiles	Developing biodegradable and eco-friendly nanomaterials.	Applications: Sustainable fashion and biodegradable clothing. Example: Nanocellulose-based fibers derived from agricultural waste.

Despite its great potentials, nanotechnology in textiles has its challenges. Regular means of functionalizing fabrics rarely yield permanent results, and the functions reduce efficiency after laundering or prolonged usage. This problem is solved by the application of nanotechnology and the incorporation of nanomaterials into the textile framework, subsequently enabling fabrics with longer durability. Care has also to be taken during manufacture in ensuring that the nanomaterials will not compromise the comfort, weight, or aesthetics of the clothing being manufactured.

Aspect	Challenges	Possible solution
Durability and wash-fastness	Ensuring nanomaterials remain functional after repeated washing and usage.	Developing strong binding mechanisms or encapsulating nanoparticles.
Health and safety concerns	Potential toxicity of certain nanomaterials, such as silver or carbon nanotubes, during manufacturing or disposal.	Conducting thorough risk assessments and adhering to safety regulations.
Cost of production	High costs associated with nanomaterial synthesis, functionalization, and integration into textiles.	Scaling up production and developing cost-effective synthesis techniques.
Environmental impact	Long-term environmental implications of releasing nanoparticles during textile production or disposal.	Promoting recyclable and biodegradable nanomaterials
Compatibility with textiles	Achieving uniform dispersion and integration of nanoparticles into different types of fibers.	Using surface modification techniques or advanced coating methods.
Regulatory challenges	Lack of standardized guidelines for nanotechnology applications in textiles.	Establishing global standards and certifications for nanotech-based textiles.
Scaling for mass production	Transitioning from laboratory-scale innovations to large-scale manufacturing.	Collaborating with textile industries to refine processes
Consumer acceptance	Concerns about the safety and added value of nanotechnology-enhanced textiles.	Educating consumers and highlighting tangible benefits.

Concerning environmental and safety issues regarding the incorporation of nanoparticles in textiles, ensuring research on the long-term effects of nanomaterials on health or ecosystems is necessary to limit any risks. There are other serious hurdles posing serious challenges to big production and commercial aspects of using nanoengineered textiles [4].

Advent of nanotechnology indicates several redefining attributes for textiles, in essence creating a universal interface. With help of integration of sensor, data transmission unit, and processing system, textiles could translate physical, chemical, electrical, and thermal information. All this is opening up new systems for smart textiles or wearables and more, forecasting a future when clothing will do much more than wear and protects; it will interface and communicate. Nanotechnology integration into textiles is a textbook case of an interdisciplinary science full of great promise. The range of possibilities that nanotechnology opens up in textiles through overcoming the limitations of textiles using traditional materials, and introducing smart textiles into the areas of fashion and design, health, defense, and numerous other sectors, redefines its future. With each passing moment leading to a progressive development in this field via continuing research, nanotechnology and textiles indicate a perfect future of wearable functionality, comfort, and sustainability. The starting point of nanotechnology into textiles is a journey of immense reputation waiting to echo for years across other industries and everyday life.

7 Conclusion

Nanotechnology enhances the properties of textiles such as durability, water repellency, antimicrobial effects, and UV protection. This makes products last longer and reduces environmental impact through reduced washing and disposal. This relates to sustainability integration into textile production, wherein the use of eco-friendly nanomaterials and green synthesis methods is encouraged. This entails developing biodegradable materials as well as practicing recycling, with principles of circular economy. Nanotechnology will be integrated in the creation of smart textiles that will monitor environmental conditions and physiological parameters to offer innovative applications in various sectors, including healthcare, sports, and defense. This necessitates further research for the challenges presented by nanomaterials usage, including health and safety risks, durability post-washing, and the environment impact of nanoparticles. Standard guidelines and safety protocols must be established to ensure that nanotechnology can be applied responsibly in textiles. Further research into new nanomaterials and their applications can lead to breakthroughs in functionality and sustainability. Research should focus on enhancing the compatibility of nanomaterials with various textile fibers and improving the cost-effectiveness of production processes.

The education of consumers about the benefits and safety of nano-enhanced textiles is important to promote acceptance and demand for these innovative products. Transparency in the production process and clear communication of the advantages of sustainable nanotextiles can drive market growth.

References

1. Rahaman, M. T., & Khan, M. S. H. (2024). Applications of green nano textile materials for environmental sustainability and functional performance: Past, present and future perspectives. *Nano-Structures & Nano-Objects, 40*, 101332.
2. Shah, M. A., Pirzada, B. M., Price, G., Shibiru, A. L., & Qurashi, A. (2022). Applications of nanotechnology in smart textile industry: A critical review. *Journal of Advanced Research, 38*, 55–75.
3. Krifa, M., & Prichard, C. (2020). Nanotechnology in textile and apparel research—An overview of technologies and processes. *The Journal of The Textile Institut.*
4. Mazari, S. A., Mubarak, N. M., Jatoi, A. S., Abro, R., Shah, A., Shah, A. K., Sabzoi, N., Baloch, H., Kumar, V., & Lghari, Z. (2021). Environmental impact of using nanomaterials in textiles. In A. K. Haghi, A. E. Kandjani, S. Thomas, & A. S. Kashyap (Eds.), *Nanosensors and nanodevices for smart multifunctional textiles* (pp. 321–342). Elsevier.
5. Omietimi, H. B., Afolalu, S. A., Kayode, J. F., Monye, S. I., Lawal, S. L., & Emetere, M. E. (2023). An overview of nanotechnology and its application. In *Proceedings of the 4th International Conference on Design and Manufacturing Aspects for Sustainable Energy (ICMED-ICMPC 2023).*
6. Asif, A. K. M. A. H., & Hasan, M. Z. (2018). Application of nanotechnology in modern textiles: A review. *International Journal of Current Engineering and Technology, 8*(2), 227–232.

7. Verma, A., Arif, R., & Jadoun, S. (2021). Impact of nanotechnology on sustainable textile material and its application. In *Green chemistry for sustainable textiles: Modern design and approaches* (pp. 165–172). Elsevier.
8. Islam, S., & ul, & Butola, B. S. (Eds.). (2018). *Nanomaterials in the wet processing of textiles*. Wiley.
9. Barman, J., Tirkey, A., Batra, S., Paul, A. A., Panda, K., Deka, R., & Babu, P. J. (2022). The role of nanotechnology-based wearable electronic textiles in biomedical and healthcare applications. *Materials Today Communications, 32*, 104055.
10. Pakdel, E., Daoud, W. A., Afrin, T., et al. (2017). Enhanced antimicrobial coating on cotton and its impact on UV protection and physical characteristics. *Cellulose, 24*, 4003–4015.

Nanotextiles for Sustainability and Circular Economy

Rey Y. Capangpangan, Arnold C. Alguno, and Anthony Halog

Abstract This chapter *"Nanotextiles for Sustainability and Circular Economy"* delves into how nanotechnology has been bringing about a significant change in the paradigms of sustainability in the textile industry which are also in tune with the principles of the circular economy. In addition, it presents models for environmentally conscious production and discusses future trends in nanotextile waste. This chapter discusses the benefits of circular design approaches—closed-loop production and upcycling of materials—to reduce waste and increase the use of renewable resources. Examples of case studies demonstrate the translation of nano-enabling sustainable solutions in industries and how regulatory issues and consumer acceptability on these novel solutions are pursued. It is envisaged in this chapter that while these technologies remain innovative and that, while sustainable practices are developed, in the future itself, nanotextiles would bring the shift to the regenerative and eco-friendly textile industry.

Keywords Nanotextiles · Life cycle analysis · Circular economy · Nanomaterials

R. Y. Capangpangan (✉)
Research on Environment and Nanotechnology Laboratories, Mindanao State University at Naawan, Naawan, Misamis Oriental, Philippines

College of Fisheries and Marine Sciences, Mindanao State University at Naawan, Naawan, Philippines
e-mail: rey.capangpangan@msunaawan.edu.ph

A. C. Alguno
Mindanao State University-Iligan Institute of Technology, Iligan City, Philippines
e-mail: arnold.alguno@g.msuiit.edu.ph

A. Halog
School of the Environment, The University of Queensland, St. Lucia, QLD, Australia
e-mail: a.halog@uq.edu.au

S. S. Muthu (ed.), *Sustainable Nano Textile Technology*, Textile Science and Clothing Technology, https://doi.org/10.1007/978-3-032-05902-4_4

1 Introduction

The textile industry, which has traditionally received praise for its creativity and economic value, now faces rising environmental concerns about its substantial ecological impact [28]. The traditional linear textile production model, which includes *take*, *make*, and *discard*, has become unsustainable because it requires large amounts of water and energy while releasing chemical pollutants and generating substantial waste [53]. The textile industry's environmental and social problems worsen because of its fast fashion cycles and synthetic fiber microplastic pollution and its inefficient recycling systems [67]. The textile sector needs to transform through new technologies and business models that focus on environmental stewardship and resource efficiency and long-term value creation to meet the world's demand for sustainable pathways [2, 64].

The textile industry has experienced a transformative shift through the development of nanotechnology [27, 45]. The manipulation of materials at nanoscale dimensions enables revolutionary innovations which produce fabrics with self-cleaning properties and antimicrobial protection and enhanced durability and smart external stimulus response capabilities. Nano-enabled textiles provide solutions to major sustainability issues that the industry currently faces while delivering new performance capabilities. Nanotechnology serves as a vital sustainability tool for the textile industry because it enables reduced production water and chemical and energy consumption and longer product lifespans and new recyclable materials [23, 43].

The achievement of full potential for nanotextile development requires alignment with circular economy principles which establish a regenerative system that reduces waste while maintaining product and material usage and natural system regeneration. The circular economy approach surpasses basic green textile production because it requires designers to rethink all aspects of product design and manufacturing and usage and disposal methods. Nanotextiles integrated into circular systems enable new approaches for closed-loop operations and modular design and extended product lifecycles and responsible material recovery, which ensure nanotechnology benefits sustainability goals without creating additional risks [55, 59].

This chapter tackles how nanotechnology combines with textile innovation to support circular economy principles. Specifically, it will (1) describe the primary sustainability issues affecting the conventional textile sector; (2) describe nanotextile basics and recent developments; (3) demonstrate how nanotextiles enable circular economy strategies through practical examples and innovative approaches; (4) examine the ethical and environmental and health implications of nano-enhanced fabrics; and (5) outline responsible research directions and design approaches and industrial implementation strategies for the future. This chapter combines technological knowledge with sustainability principles to provide a thorough understanding of nanotextiles as drivers for a circular and regenerative textile industry future.

2 The Environmental Impact of Conventional Textiles

The textile industry maintains its position as a fundamental economic and social pillar of global economies but it remains one of the most resource-intensive and environmentally damaging sectors. The conventional production systems of the industry depend on virgin resource extraction and energy-intensive manufacturing and linear consumption patterns which produce extensive ecological and social impacts. The extent of these impacts needs thorough assessment to drive innovation toward sustainable practices [19, 21].

2.1 *Resource Consumption: Water, Energy, and Raw Materials*

Textile production using conventional methods requires enormous amounts of water together with substantial energy consumption and large quantities of raw materials. The fiber cotton stands out as one of the most widely used materials because it creates an enormous water consumption during its production process. The water requirements for producing one kilogram of cotton amount to between 10,000 and 20,000 liters based on farming conditions [12]. The use of water in irrigation practices results in the depletion of freshwater resources and simultaneously causes damage to land and destruction of biodiversity.

Textile manufacturing requires substantial amounts of energy throughout its stages including fiber production dyeing finishing and laundering [42, 49]. The majority of textile production depends on fossil fuels as its energy source which leads to substantial greenhouse gas emissions. The Ellen MacArthur Foundation [16] reports that textile manufacturing produces 1.2 billion tonnes of CO_2 equivalent annually which surpasses the combined emissions from international flights and maritime shipping.

The use of raw materials especially synthetic fibers made from petroleum such as polyester increases environmental pressures. The production of polyester requires large amounts of energy and results in higher carbon emissions than the traditional methods of cotton production [62]. The extraction and processing of non-renewable resources cause ecological damage and reduce the availability of resources [39, 54].

2.2 *Pollution and Waste Generation*

The traditional textile industry produces significant environmental pollution. Textile dyeing and treatment operations use extensive amounts of chemicals which result in 17%–20% of global industrial water pollution according to Kant [29]. The chemicals used in this industry prove toxic while being non-biodegradable and they cause harm to both aquatic ecosystems and human health.

The increasing problem of microplastic pollution has become a major environmental concern. Synthetic fabrics release microplastics during washing which then migrate into water systems and ocean environments. Research indicates that washing a single load of clothing produces hundreds of thousands of microfibers, which substantially adds to the marine microplastic crisis [11].

The industry produces massive quantities of solid waste as part of its operations. The fast fashion model drives fast consumption and quick disposal of clothing which produces millions of tons of textile waste throughout the year [3, 10]. According to the Ellen MacArthur Foundation [16], the textile recycling rate for new clothing production stands at less than 1% while most textiles end up in landfills or get incinerated.

2.3 *Life Cycle Assessment (LCA) of Traditional Textiles*

The Life Cycle Assessment (LCA) methodologies offer a complete understanding of textile environmental impacts through all stages from raw material extraction to the final disposal. The different stages of a textile product's existence generate substantial environmental damage according to LCAs [31].

The use phase which includes washing and drying and ironing of garments represents a major portion of total energy usage and carbon emissions according to Beton et al. [8]. The disposal phase of products results in waste production while failing to achieve meaningful material recovery. Synthetic fibers decompose slowly in landfills because of greenhouse gas methane emissions while incineration produces carbon dioxide and toxic pollutants.

Research based on LCA demonstrates that synthetic fibers such as polyester produce less water usage than cotton but create more GHG emissions and microplastic pollution according to Shen et al. [62]. Systemic thinking emerges as a necessary approach to tackle textile sustainability challenges.

2.4 *The Need for Innovation Toward Sustainability*

The industry needs immediate radical innovation to address the extensive environmental consequences of conventional textiles. The industry needs more than minor improvements in dyeing processes to achieve sustainability.

The path to sustainability requires sustainable innovation to address various fronts:

- Material innovation, including bio-based and recyclable fibers.
- Process optimization, such as waterless dyeing technologies.
- Circular business models, emphasizing reuse, repair, recycling, and responsible design.
- Consumer engagement, encouraging sustainable purchasing and use behaviors.

Nanotechnology provides promising solutions through its ability to create fabrics which need less washing and resist stains and last longer thus decreasing resource usage throughout their entire lifecycle. The deployment of these innovations through circular economy frameworks will maximize environmental benefits while preventing new risks from emerging [66].

3 Understanding Nanotextiles

3.1 Definition and Key Characteristics of Nanotextiles

Nanotextiles describe textile materials that receive advanced functionalities through nanoscale engineering (1–100 nm) modifications. The integration of nanotechnology into fibers and fabrics and finishes enables scientists to modify material properties at the atomic or molecular level which results in new capabilities without weight or thickness changes [13].

The main features of nanotextiles include:

- *Enhanced performance*: Greater strength, flexibility, and durability compared to conventional textiles.
- *Multi-functionality*: Ability to combine multiple features (e.g., water repellence, antimicrobial action) in a single fabric.
- *Lightweight nature*: Minimal addition of mass or volume despite significant functional enhancements.
- *Long-term stability*: Improved resistance to environmental factors like UV radiation, moisture, and mechanical wear.
- *Responsiveness*: Some nanotextiles can react to external stimuli such as temperature, pH, light, or moisture, demonstrating "smart" behavior [37].

3.2 Common Nanomaterials Used in Nanotextiles

Nanotextiles are made from various types of nanomaterial, each of which confers different properties for specific applications:

(a) Nanoparticles.

- Silver nanoparticles (AgNPs): are a widely utilized antibacterial and antifungal agent in many fields [46].
- Titanium dioxide (TiO_2) nanoparticles: exhibit photocatalytic activity leading to self-cleaning and ultraviolet (UV) protection properties [51].
- Zinc oxide (ZnO) nanoparticles: are employed in coatings and fiber systems, work as UV protectants and antimicrobial agents [26].

(b) Nanofibers.

- The process of electrospinning yields materials of high surface area-to-volume ratios.
- Such nanofibers made from PVA, PAN, and chitosan polymers find applications in the textile industry such as the manufacture of filtration system, moisture management textiles, and wound dressing materials [75].

(c) Nanocoatings.

- Nanoscopic thin film is coated onto the textile surface by sol-gel process, plasma treatment, and layer-by-layer assembly approach [17].
- Water repellency and stain release and chemical protective properties are also possessed by nanocoatings which do not compromise the breathability and softness of fabric [44].

3.3 Functional Enhancements Provided by Nanotechnology

There are multiple functional benefits in incorporating nanotechnology for textile that surpass the conventional textile capabilities:

(a) Durability.

Applying nanotechnology to fabrics leads to superior strength material that helps in less wear over the period of time. Wang et al. [68] report on the significant enhancement of tensile strength and ductility of fibers through the inclusion of carbon nanotubes or graphene.

(b) Self-Cleaning Properties.

The imitation of the lotus effect gives rise to the development of nanotextiles deposited with superhydrophobic coatings (TiO_2 or SiO_2 nanoparticles) that can have self-cleaning surfaces. The nano/microstructure of these fabrics prevents adhesion of water and soiling, reducing the need for both frequent washing and detergent [63].

(c) Antibacterial Properties.

Silver nanoparticles (AgNPs) are particularly effective in eliminating various bacteria. Proceeding via antimicrobial action against membranes and interference with respiratory enzyme, AgNP-functionalized fabrics prevent odors and infection, making it ideal for sportswear, medical textiles, and personal protection [46].

(d) UV Resistance.

Fabric doped with nanoparticles such as TiO_2 and ZnO has been introduced for protection from ultraviolet (UV) radiation. These particles will absorb and scatter harmful UV rays protecting the wearer's skin and adding life to the fabric by reducing photo degradation [22].

(e) Smart and Responsive Behaviors.

Novel nanotextiles exhibit the ability to respond to environmental conditions:

- Phase change materials (PCMs) or stimuli-responsive polymers can change their thermal properties with temperature, offering shape-stabilized temperature control [58].
- Through rapid sweat transport, moisture management is achieved with nanofiber membrane that keeps users dry and comfortable [69].
- By integrating conductive nanoparticles or fibers in textiles, textiles can detect both physiological signals (temperature, pressure, etc.) and bio-signals [56], resulting in smart clothing and on-body health monitoring systems.

4 Sustainability Advantages of Nanotextiles

The implementation of nanotechnology in textiles brings both performance improvements and major environmental advantages. Nanotextiles provide promising answers to several sustainability problems that conventional textiles face regarding resource consumption and product durability and waste reduction and environmentally friendly material development. Nanotextiles support environmental stewardship principles and circular economy thinking to transform the textile industry into a sustainable future direction.

4.1 Resource Efficiency: Less Water, Chemicals, and Energy Use

The main sustainability benefit of nanotextiles emerges from their resource-efficient production methods. The production of traditional textiles requires substantial amounts of water and chemicals during dyeing and finishing operations. Nanocoatings and nanofinishing techniques provide a solution to minimize water consumption and chemical reagent usage and energy requirements.

The sol-gel process for functional nanocoating applications requires less water and chemical additives than traditional wet-processing techniques according to Asuka & Sigmund [4]. Plasma-enhanced treatments enable nanoscale functionalization of textiles through methods that eliminate the requirement of large chemical auxiliaries and water consumption thus producing cleaner and more energy-efficient outcomes [60].

The integration of self-cleaning nanostructures containing TiO_2 or SiO_2 nanoparticles into fabrics enables reduced washing needs which decreases household water and energy usage throughout the product's entire lifespan [48]. The efficient use of resources through this method minimizes the environmental impact that occurs during textile use.

4.2 Extended Product Life: Durability and Reusability

Nanotextiles receive engineering to enhance their durability through resistance against mechanical wear and chemical degradation and environmental stressors including UV exposure and moisture. The enhanced durability of textile products through these improvements directly addresses the main sustainability issue related to fast fashion and short-lived garments.

Zinc oxide and silica nanoparticles applied to fabrics create enhanced abrasion resistance and UV protection which preserves both structural integrity and visual appeal throughout extended usage periods [6]. The antibacterial properties of silver nanoparticles extend product usability because they prevent odors from forming which reduces the need for frequent washing [46].

Nanotextiles achieve waste reduction through their ability to last longer while reducing the need for new resource-intensive manufacturing which aligns with circular economy principles of reuse and longevity.

4.3 Reduced Waste and Emissions

Nanotextiles deliver improved functionality together with extended operational periods which results in major reductions of textile waste. The Ellen MacArthur Foundation [16] reports that extending clothing lifespan by 9 months decreases carbon and water consumption and waste output by 20%–30% for each garment. Nanotextiles are engineered to reach these extended product lifecycles.

The self-cleaning and antimicrobial features of nanotechnology-based materials reduce the requirement for regular washing and early product disposal because of stains or odors or material breakdown. The reduction of greenhouse gas emissions occurs because of lower emissions during both production and maintenance phases [15].

The production waste from nanofiber technologies remains low because these technologies use controlled material amounts and produce fewer offcuts and defects than traditional spinning and weaving methods. The precise engineering of nanofiber mats and coatings enables manufacturers to minimize their manufacturing waste output [33].

4.4 Potential for Biodegradability and Eco-Design

The development of nanotextiles now focuses on bio-based nanomaterials and green nanotechnologies which promote biodegradability and eco-design. Synthetic nanomaterials have traditionally faced criticism because of their potential to persist in the environment. The development of biodegradable nanocomposites together with natural-origin nanofibers now tackles the existing environmental risks.

Scientists investigate chitosan and cellulose nanocrystals and starch-based nanostructures to develop nano-enhanced textiles which break down safely after use without toxic substance release [74]. The new technologies follow eco-design principles because they start with environmental compatibility in mind for the complete product life cycle.

The precise engineering of nanotextiles enables the creation of mono-material structures which simplifies end-of-life recycling and material recovery operations for a sustainable circular economy [16].

5 Nanotextiles and Circular Economy: A Strategic Fit

The textile industry needs to adopt a circular economy transformation which moves away from linear "take-make-dispose" operations toward a regenerative system that reduces waste while optimizing resource use. Nanotextiles possess unique capabilities that make them ideal for advancing the industry shift toward sustainability. Nanotechnology enables sustainable textile ecosystems through innovative approaches to product design and use and end-of-life management.

5.1 Principles of Circular Economy Applied to Textiles

The circular economy functions through three essential principles which Ellen MacArthur Foundation established in 2013:

- Design out waste and pollution: The initial design phase of products aims to reduce their negative environmental effects.
- Keep products and materials in use: The strategies of reuse and repair along with remanufacture and recycling help maintain product and material utilization.
- Regenerate natural systems: The system should maintain natural cycles while preserving finite resources from depletion.

The principles of circular economy apply to textiles through the following practices:

- Circular design: The creation of durable fabrics and garments should include modular designs which enable easy disassembly and recycling as well as repairability and modularity.
- Reuse and remanufacturing: The product lifespan can be extended by creating reuse markets and through upcycling and reconditioning processes.
- Recycling: The system enables complete fiber-to-fiber recycling which produces high-quality materials while maintaining their original quality.

5.2 How Nanotextiles Support Circularity

(a) Enabling Closed-Loop Material Cycles.

The application of nano-enabled durability and performance enhancements directly supports closed-loop material systems through their ability to decrease textile fiber deterioration during use and recycling operations. Nanotechnology develops protective coatings and fiber structures which preserve material strength through repeated use and recycling operations thus reducing material deterioration.

For example, graphene-based textile modifications may offer a prospect to enhance fiber strength and conductivity without substantially affecting recyclability [32]. This allows textiles to be recycled several times while retaining functionality, which is essential for closed-loop systems.

(b) Facilitating Modular and Adaptable Textile Design.

The modular design approach enables easy disassembly and update of garments which makes repair and remanufacturing and recycling processes more accessible. Nanotextiles support this approach through:

- Smart coatings that can be selectively activated or removed.
- Fibers that respond to stimuli can modify their properties such as color and texture without needing fresh materials [9].

Designers can use these capabilities to develop adaptable clothing items which extend product longevity and decrease the need for continuous new purchases.

(c) Enhancing Recyclability and Material Recovery.

Textile recycling faces a significant challenge because of the intricate nature of material blends which include polyester-cotton mixes. The combination of nano-engineered mono-material solutions with functional nanocoatings enables the preservation of recyclability while delivering performance characteristics that blends typically provide.

Nanotechnology facilitates such tracking of materials by the introduction of nano-sized labels into fibers, which provide the capability of automatic sorting during recycling [25]. The integration of these technologies will significantly improve the efficiency and scalability of textile recycling systems and will solve a major barrier against circularity.

5.3 Case Studies of Circular Textile Initiatives

Several initiatives illustrate how innovations aligned with circular economy principles are being implemented in the textile sector:

(a) PANGAIA: Sustainable Materials Enhanced by Nanotechnology.
The material science brand PANGAIA implements nanotechnology to boost the performance and longevity of sustainable textile materials. The FLWRDWN material from PANGAIA uses natural wildflowers with biopolymer nanotechnology to make biodegradable high-performance thermal insulation that supports composting at the end of its life cycle [50].
(b) Worn Again technologies: Fiber-to-Fiber Recycling.
Worn Again Technologies operates chemical recycling systems that extract polyester and cellulose from blended fabrics for complete textile recycling although their main focus is not on nanotextiles. Nano-enabled markers would enhance sorting and recovery operations in such systems according to Worn Again Technologies [71].
(c) Nano-enhanced Smart labels for Recycling.
The RESYNTEX project funded by the European Union investigates nano-enabled smart labels which store material details to enhance automated disassembly and material recovery operations at recycling facilities [57].

6 Innovations and Applications in Sustainable Nanotextiles

The advancement of sustainable nanotextiles creates an innovative boundary between nanotechnology and critical environmental requirements and social needs. The textile industry now provides sustainable products that perform well and use resources efficiently and protect the environment through bio-based nanomaterials and eco-friendly nanocoatings and smart functionalities. Sustainable nanotextile applications show extensive transformative potential through innovations that span fashion textiles and medical textiles and technical fabrics and sportswear.

6.1 Bio-Based Nanomaterials for Textiles

The shift toward sustainable nanotextiles continues to progress as more manufacturers turn to bio-based (instead of petro-based) nanomaterials. The natural materials are highly functional and biodegrade in a manner that does not harm the environment.

Examples of bio-based nanomaterials include:

- CNCs derived from plant fibers exhibit strength/lack of stiffness properties which make them applicable in enhancing textiles as wrinkle resistance and durability improvement treatments [18].
- Antimicrobial activity of chitosan nanoparticles derived from crustacean sloughing would make these applicable for textile uses in treatment of wounds and in hygiene purposes [36].

- Biodegradable and renewable starch and soy-based nanomaterials are currently being investigated for their application as sustainable fabric coating and structural materials [34].

These materials provide textiles with high-performance attributes and the possibility of biodegrading at the end of the product life, for a circular economy approach.

6.2 Eco-Friendly Nanocoating and Nanofinishes

Conventional textile finishes generally include large quantities of noxious chemicals and solvents. On the other hand, eco-friendly nanocoatings pursue a similar or even better performance, such as water repellency, stain resistant, and UV protection, by minimizing environment harm.

Examples include:

- Water-based sol-gel processed silica-based hydrophobic coatings which can provide self-cleaning and water-repellency with non-emitting toxic fumes [73].
- Silver nanoparticles synthesized using plant extracts (green synthetic protocols) reducing adverse environmental impacts as well as providing antimicrobial activity [1].

These eco-friendly finishes promote a superior path of sustainable production by decreasing dependence on unsafe chemicals and allowing for lower temperature, less energy-requiring finishing methods.

6.3 Smart Textiles for Energy Harvesting and Environmental Monitoring

Due to the development of functional nanomaterials, smart nanotextiles are meaningfully developed for the purposes of energy harvesting and environmental monitoring, thereby opening up new horizons for textiles:

- Energy harvesting: Fabrics with electrically active piezoelectric nanothreads (e.g., zinc oxide nanowires) have been created and produce current when mechanically deformed, such as when people walk or adjust their bodies [5]. This breakthrough could enable self-powered systems for, wearable electronics, rackless light displays, and ornamental watchbands.
- Environmental monitoring: The integration of conductive nanomaterials, (e.g., carbon nanotubes, graphene) into textiles allows to the real-time sensing of pollutants, temperature, humidity, biochemical markers [14]. These are character-

ized to be very useful in environmental monitoring, health diagnostics, and industrial safety.

These intelligent qualities not only bring convenience to users but also contribute to wider-spread sustainability. They let users to be energy independent and self-aware of the genius of nature.

6.4 Applications across Industries

(a) Fashion.
The fashion industry produces self-cleaning garments with anti-odor properties and UV protection through nanotextiles which extend the life of clothing by reducing washing needs. PANGAIA uses bio-based nanotechnologies to develop sustainable clothing products that maintain high-performance standards [50].

(b) Medical Textiles.
Medical textiles receive nanoenhancements to create wound dressings with antimicrobial properties and smart bandages which track healing advancement and surgical gowns with protective functions. Silver nanoparticle-infused dressings serve to stop infections while helping patients recover more quickly [46].

(c) Technical Fabrics.
The technical sector uses protective clothing which serves military personnel as well as industrial workers and firefighters. Nano-enhanced textiles provide better thermal insulation and chemical defense capabilities and sensor functionality which enhances protective clothing performance in harsh environments [70].

(d) Sportswear.
Nanotextiles such as in sportswear offer enhanced moisture absorption, transpiration, as well as antibacterial with durability. Under Armour and Nike both use nanofinishes to produce lightweight sportswear which retains its performance over several wear cycles [41].

6.5 Success Stories and Industrial Practices

- HeiQ Materials AG leads textile innovation through its development of bluesign and OEKO-TEX approved silver and peppermint oil-based antimicrobial technologies which support both performance and sustainability [20].
- The nanotechnology developed by W. L. Gore & Associates produces Gore-Tex fabrics with waterproof breathable membranes that are transitioning to PFC-free (perfluorinated compound-free) versions to decrease toxic chemical consumption [38].

- The Water<Less™ Process of Levi's® uses innovative finishing methods to minimize water consumption in denim manufacturing while future nanotechnology applications could enhance these sustainable processes (*Life Cycle Assessment as a Next Level of Transparency in Denim Manufacturing*, 2023).

The examples show sustainable nanotextiles have moved beyond laboratory testing because they now enter mainstream markets while demonstrating scalability and commercial viability and environmental benefits.

7 Challenges and Ethical Considerations

Nanotextiles bring promising sustainability and innovation potential yet they create major environmental and health risks together with ethical dilemmas. The benefits of nanotechnology need to be extracted through balanced methods that reduce unintended consequences. Early intervention on these challenges is essential to guarantee safe and responsible development and deployment and disposal of nanotextiles.

7.1 Potential Environmental Risks of Nanomaterials

The small dimensions and high reactivity of nanomaterials create distinctive environmental hazards. Nano-enhanced textiles release nanoparticles into wastewater during washing and abrasion and disposal which then enter natural ecosystems [7]. Silver nanoparticles (AgNPs), which serve as antimicrobial agents, have the potential to build up in aquatic organisms while disrupting microbial communities and aquatic food webs [72].

Engineered nanomaterials (ENMs) face concerns about their ability to persist and accumulate in soil and water ecosystems. The interactions between nanomaterials and living cells remain poorly understood and scientists have not been able to measure the complete ecological effects of these materials over time [24]. The environmental discharge of nanoscale pollutants from textiles creates a critical situation which demands immediate resolution.

7.2 Health and Safety Concerns

Nanomaterials pose health risks because they can penetrate biological membranes and enter cells while causing oxidative stress and inflammation and genotoxic effects [65]. The potential exposure to nanoparticles exists for textile manufacturing workers and consumers of nano-enhanced garments and waste handlers through inhalation and dermal contact and ingestion.

It has been shown in animal models that inhalation of carbon nanotubes (CNT) and titanium dioxide (TiO2) nanoparticles result in pulmonary inflammation and fibrosis [61]. There continues to be a concern about the risks associated with textile exposure because of the possible cumulative harm from repeated exposure to the people.

The absorption of nanoparticles through skin during extended nano-treated garment use has become a rising concern although research indicates that most nanomaterials do not penetrate skin under typical conditions [35]. Systematic human exposure studies that follow participants over time are currently absent from the scientific literature.

7.3 Issues in Recycling Nano-Enhanced Textiles

The recycling process of nanotextiles creates specific difficulties:

- Nano-enhanced fabrics present complex material structures because they contain multiple material components including natural fibers with nanocoatings and embedded nanoparticles, which makes mechanical and chemical recycling operations more difficult.
- The introduction of nanoparticles into recycling operations creates a risk of contaminating new materials which could result in uncontrolled nanomaterial distribution throughout secondary products [76].
- Nanomaterials show unpredictable behavior during recycling operations because they break down into smaller fragments or form aggregates which impacts both material quality and environmental safety.

The current recycling technologies lack proper detection and separation capabilities for nano-enabled components, which prevents true circularity in nanotextile recycling.

7.4 Regulatory and Standardization Gaps

The fast commercialization of nanotextiles has not been accompanied by adequate regulatory frameworks that address nanomaterials in textiles:

- The European Chemicals Agency reports that the Registration, Evaluation, Authorisation, and Restriction of Chemicals (REACH) regulation requires nanomaterial registration but textile-specific guidelines are still developing [40].

- The United States Environmental Protection Agency (EPA) established reporting requirements for nanomaterials under the Toxic Substances Control Act (TSCA) but the enforcement and evaluation standards are not consistent.
- The OECD [47] reports that there are no standardized global standards for measuring nanoparticle release from textiles or for evaluating environmental risks or providing nano-enabled product labels for consumer awareness.

The absence of standardized global standards creates potential risks for both industrial progress and public confidence.

7.5 Strategies for Safe and Responsible Innovation

The following set of strategies should be adopted to reduce nanotextile risks:

(a) Safe-by-Design Approaches.
    The early implementation of safety measures through nanomaterial selection and encapsulation methods and recyclable design reduces environmental and health risks [30].
(b) Life Cycle Assessment (LCA) Including Nanomaterials.
    The entire life cycle of nanoproducts can be analyzed through nano-specific life cycle assessments which predict environmental effects from raw material extraction to disposal [52]. The actual sustainability metrics enable developers to evaluate nanotextiles relative to traditional alternatives.
(c) Transparency and Consumer Communication.
    The combination of nano-enabled material disclosure with material composition information enables consumers to make informed decisions while making industries accountable for their actions.
(d) Robust Testing and Monitoring Frameworks.
    Standardized testing protocols for nanoparticle release and toxicity assessments and exposure monitoring will improve regulatory oversight and enable safe market entry of new products [47].
    The textile industry can implement nanotechnology safely through the integration of precautionary principles with ethical responsibility and interdisciplinary collaboration to protect human and environmental health.

8 Future Perspectives

Future success in sustainable nanotextiles requires strategic efforts to link scientific innovation with systemic change for complete realization of its potential. Bio-nanotechnology and green synthesis methods need to become the primary focus because they reduce environmental risks while delivering maximum performance and renewability in nanomaterial development. Digital technologies such as blockchain need integration throughout the textile supply chain to track products while enhancing recycling operations and verifying sustainable practices which strengthens circular economy systems. The establishment of stronger policy frameworks must begin to effectively regulate nano-enabled textiles and education initiatives should teach sustainability principles to develop future leaders and innovators. The approach to consumer engagement needs transformation because it should encourage people to value products that last longer and can be recycled and have transparent manufacturing processes. The research needs to focus on developing nanotextiles that are completely biodegradable and scaling eco-friendly manufacturing at industrial levels and advancing smart recyclable textiles and creating standardized nano-specific life cycle assessments to guide responsible design and production. Stakeholders who take proactive and integrative steps will establish nanotextiles as foundational elements for achieving technological excellence and environmental stewardship in the evolving global economy.

9 Conclusion

This chapter demonstrates how conventional textiles create severe environmental problems yet nanotextiles offer solutions through their ability to improve resource efficiency and extend product life and reduce waste and enable eco-design innovations. Nanotextiles provide sustainable and circular textile economy solutions through their use of bio-based nanomaterials and green synthesis methods and smart functionalities and digital traceability tools. A responsible and holistic approach must be adopted to achieve this vision because it requires understanding the environmental and health and ethical risks of nanomaterials while building safety and transparency and recyclability into initial designs. Sustainable nanotextiles will succeed in the future through combined research efforts between scientists who develop environmentally conscious innovations and industrial sectors that implement circular design principles and governmental entities that create science-based regulatory frameworks and incentives for responsible adoption. The stakeholders working together will establish an integrated system which combines technological innovation with planetary health objectives and resource regeneration and societal well-being.

References

1. Adaikalam, A., & Arumugam, M. (2024). Eco-friendly synthesis of silver nanoparticles using Corymbia ptychocarpa extracts: Characterization and antibacterial activity. Nanotechnology Perceptions, 298–307. https://doi.org/10.62441/nano-ntp.vi.3479
2. Annapoorani, S. (2023). Reality and challenges in sustainable textiles (pp. 47–72). Springer International Publishing. https://doi.org/10.1007/978-3-031-35451-9_3

3. Arias-Meza, M., Alvarez-Risco, A., Cuya-Velásquez, B. B., de las Mercedes Anderson-Seminario, M., & Del-Aguila-Arcentales, S. (2022). Fashion and textile circularity and waste footprint. In A. Alvarez-Risco, S. S. Muthu, & S. Del-Aguila-Arcentales (Eds.), Circular economy. Environmental footprints and eco-design of products and processes. Springer. https://doi.org/10.1007/978-981-19-0549-0_9
4. Asuka, M., & Sigmund, W. M. (2008). Low temperature transparent nanoplatelet hybrid gas barrier coating materials via sol-gel process. MRS Proceedings, 1114(1), 50401–50409. https://doi.org/10.1557/PROC-1114-G05-04
5. Badawy, D. A., Yang, H., Ko, F., & Wang, L. (2014). Piezoelectric zinc oxide woven fabric sensors for dynamic force and damage sensing. https://www.dpi-proceedings.com/index.php/dmac2014/article/view/21311.
6. Becheri, A., Dürr, M., Lo Nostro, P., & Baglioni, P. (2008). Synthesis and characterization of zinc oxide nanoparticles: Application to textiles as UV-absorbers. Journal of Nanoparticle Research, 10(4), 679–689. https://doi.org/10.1007/S11051-007-9318-3
7. Beigzadeh, Z., Kolahdouzi, M., Kalantary, S., & Golbabaei, F. (2024). A systematic review of released nano-particles from commercial nano-textiles during use and washing. Journal of Industrial Textiles, 54. https://doi.org/10.1177/15280837241254512
8. Beton, A., Dias, D., Farrant, L., Gibon, T., Desaxce, M., Perwueltz, A., & Boufateh, I. (2014). Environmental improvement potential of textiles. (IMPRO Textiles). European Commission Joint Research Centre.
9. Bhat, A. H., Khan, I., Bhawani, S. A., Rahim, M. K. A., & Gazal, U. (2021). Stimuli-responsive polymer composites for fabric applications (pp. 351–376). Woodhead Publishing. https://doi.org/10.1016/B978-0-12-819961-9.00018-9
10. Bick, R., Halsey, E., & Ekenga, C. C. (2018). The global environmental injustice of fast fashion. Environmental Health, 17(1), 1–4. https://doi.org/10.1186/S12940-018-0433-7
11. Browne, M. A., Crump, P., Niven, S. J., Teuten, E., Tonkin, A., Galloway, T., & Thompson, R. (2011). Accumulation of microplastic on shorelines worldwide: Sources and sinks. Environmental Science & Technology, 45(21), 9175–9179. https://doi.org/10.1021/es201811s
12. Chapagain, A., Hoekstra, A. Y., Savenije, H. H. G., & Gautam, R. (2005). The water footprint of cotton consumption (pp. 103–130). UNESCO-IHE Institute for Water Education. https://doi.org/10.1002/9780470696224.CH9
13. De Azeredo, H. M. (2009). Nanocomposites for food packaging applications. Food Research International, 42(9), 1240–1253. https://doi.org/10.1016/j.foodres.2009.03.019
14. Eghan, B., Ofori, E. A., Seidu, R. K., & Acquaye, R. (2024). Recent progress to address the challenges of conductive inks for e-textiles. Engineering Proceedings. https://doi.org/10.3390/engproc2023052003
15. El-Khatib, E. M. (2012). Antimicrobial and self-cleaning textiles using nanotechnology. Research Journal of Textile and Apparel, 16(3), 156–174. https://doi.org/10.1108/RJTA-16-03-2012-B016
16. Ellen MacArthur Foundation, A new textiles economy: Redesigning fashion's future (2017). Date Accessed: April, 29, 2025.
17. Gashti, M. P., Pakdel, E., & Alimohammadi, F. (2016). Nanotechnology-based coating techniques for smart textiles (pp. 243–268). Woodhead Publishing. https://doi.org/10.1016/B978-0-08-100263-6.00011-3
18. Ghazal, H., Maraae, A., Beltagy, Z., Shamy, M., Nasser, A., Abd-Elaal, L., & Allam, L. (2024). A review on cellulose nanocrystals (CNCs) as green finishing material to produce multifunctional textiles. Egyptian Journal of Textile and Polymer Science and Technology. https://doi.org/10.21608/jtcps.2024.259520.1302
19. Harsanto, B., Primiana, I., Sarasi, V., & Satyakti, Y. (2023). Sustainability innovation in the textile industry: A systematic review. Sustainability, 15(2), 1549. https://doi.org/10.3390/su15021549
20. HeiQ. (2020). https://plbunited.com/products-heiq-products.html Date accessed: 04/30/2025.

21. Hongal, P., & Kshirsagar, Y. (2024). Crafting a sustainable textile industry: Challenges and opportunitites from fibre to fabrics. International Journal of Advanced Research, 12(08), 1553–1562. https://doi.org/10.21474/IJAR01/19403
22. Hossain, M. A., & Rahman, M. (2015). A review of nano particle usage on textile material against ultra violet radiation. Journal of Textile Science and Technology, 01(3), 93–100. https://doi.org/10.4236/JTST.2015.13010
23. Hossain, M. M., Al-Tabbaa, O., & Ahammad, M. F. (2025). Environmental sustainability in textile and apparel global value chain: Towards achieving the United Nations sustainable development goals. In Sustainable textile and apparel chain management (pp. 7–57). Springer. https://doi.org/10.1007/978-3-031-80240-9_2
24. Hou, W. C., Hou, W. C., Westerhoff, P., & Posner, J. D. (2013). Biological accumulation of engineered nanomaterials: A review of current knowledge. Environmental Science: Processes & Impacts, 15(1), 103–122. https://doi.org/10.1039/C2EM30686G
25. Iezzi, B., Shtein, M., Wang, T., & Rothschild, M. (2024). Fiber and fabric-integrated tracing technologies for textile sorting and recycling, 223–237. https://doi.org/10.1002/9781394214297.ch18
26. Irede, E., Awoyemi, R. F., Owolabi, B., Aworinde, O. R., Kajola, R. O., Hazeez, A., Raji, A. A., Ganiyu, L. O., Onukwuli, C. O., Onivefu, A. P., & Ifijen, I. H. (2024). Cutting-edge developments in zinc oxide nanoparticles: Synthesis and applications for enhanced antimicrobial and UV protection in healthcare solutions. RSC Advances, 14(29), 20992–21034. https://doi.org/10.1039/d4ra02452d
27. Isaev, X. (2024). Advancements in nanotechnology: Revolutionizing Uzbekistan's textile industry. International Journal of Advances in Scientific Research, 4, 41. https://doi.org/10.37547/ijasr-04-04-08
28. Jovanović, M., & Urošević, S. (2024). Impact of the textile industry on the environment. In 7th International Scientific Conference on Contemporary Trends and Innovations in Textile Industry (pp. 559–567). https://doi.org/10.5937/ct_iti24062j
29. Kant, R. (2012). Textile dyeing industry an environmental hazard. Natural Science, 4(1), 22–26. https://doi.org/10.4236/ns.2012.41004
30. Keshavan, S., Petri-Fink, A., & Rothen-Rutishauser, B. (2024). Understanding the benefits and risks of sustainable nanomaterials in a research environment. Chimia, 78(6), 397–402. https://doi.org/10.2533/chimia.2024.397
31. Khan, R. M. W. U., & Shaker, K. (2023). Life cycle assessment of textile products (pp. 159–176). Springer Nature. https://doi.org/10.1007/978-3-031-49479-6_6
32. Khoso, N. A., Ahmed, A., Memon, S. I., Jamshaid, H., Rehan Abbasi, A. M., Qutaba, S., & Wang, J. (2024). Fabrication of graphene oxide and conductive polymer on cotton fabric with ultrasonic (US) assisted dyeing for washable E-textiles. https://doi.org/10.20944/preprints202406.1920.v1
33. Kim, H.-Y. (2006). Method of manufacturing mats consisting of nanofibers by electrospinning and mats manufactured thereby. https://patents.google.com/patent/WO2006123858A1/en.
34. Koshy, R. R., Mary, S. K., Pothan, L. A., & Thomas, S. (2015). Soy protein- and starch-based green composites/nanocomposites: Preparation. In Properties, and applications (pp. 433–467). Springer. https://doi.org/10.1007/978-81-322-2470-9_14
35. Leonida, M. D., & Kumar, I. (2016). Nanoparticles. In Nanomaterials and Nanocarriers (pp. 37–46). Springer. https://doi.org/10.1007/978-3-319-39168-7_5
36. Lin, X., Gong, X.-Z., Ruan, Q., Xu, W., Zhang, C., & Zhao, K. (2022). Antimicrobial application of chitosan derivatives and their nanocomposites. Current Medicinal Chemistry, 30(15), 1736–1755. https://doi.org/10.2174/0929867329666220803114729
37. Mahltig, B., & Textor, T. (2013). Nanotechnology in textiles: Concepts, properties, applications. Walter de Gruyter.
38. Maity, S. (2022). Waterproof breathable fabrics and suits (pp. 347–375). Elsevier eBooks. https://doi.org/10.1016/b978-0-323-90477-3.00015-8

39. Moazzem, S., Crossin, E., Daver, F., & Wang, L. (2018). Baseline scenario of carbon footprint of polyester T-shirt. Journal of Fiber Bioengineering and Informatics, 11(1), 1–14. https://doi.org/10.3993/JFBIM00262
40. Montesinos-Navarro, N. A. (2022). Reach. Elsevier eBooks. https://doi.org/10.1016/b978-0-12-824315-2.00232-3
41. Nakkeeran, S. (2023). Emerging Nano-enable materials in the sports industry. Materials Research Foundations, 75–100. https://doi.org/10.21741/9781644902288-4
42. Naqvi, S. M. K., Bakhtiar, A., & Mahachandra, M. (2024). Energy audit of a textile company: A case study of finishing textile. World Journal Of Advanced Research and Reviews, 22(3), 1040–1047. https://doi.org/10.30574/wjarr.2024.22.3.1538
43. Ngô, C., & Van de Voorde, M. H. (2014). Nanotechnology for the textile industry (pp. 321–329). Atlantis Press. https://doi.org/10.2991/978-94-6239-012-6_19
44. Nivetha, A., & Pottail, L. (2024). Silane-free superhydrophobic coatings from ball-milled nanosand using eco-friendly polymeric modifiers. Journal of Dispersion Science and Technology, 1–13. https://doi.org/10.1080/01932691.2024.2432987
45. Noor-Evans, F., Peters, S., & Stingelin, N. (2012). Nanotechnology innovation for future development in the textile industry (pp. 109–131). Woodhead Publishing. https://doi.org/10.1533/9780857095190.2.109
46. Nowack, B., Krug, H. F., & Height, M. (2011). 120 years of nanosilver history: Implications for policy makers. Environmental Science & Technology, 45(4), 1177–1183. https://doi.org/10.1021/es103316q
47. OECD (2019). Guiding principles for measurements and reporting for nanomaterials: Physical chemical parameters.
48. Pakdel, E., Daoud, W. A., Kashi, S., Parvinzadeh Gashti, M., & Wang, X. (2024). Superhydrophilic self-cleaning cotton fabric with enhanced antibacterial and UV protection properties. Cellulose. https://doi.org/10.1007/s10570-024-06346-1
49. Palamutcu, S. (2015). Energy footprints in the textile industry (pp. 31–61). Woodhead Publishing. https://doi.org/10.1016/B978-0-08-100169-1.00002-2
50. PANGAIA. https://pangaia.com/pages/flwrdwn?srsltid=AfmBOorzghEhTTe98J7JW6m1gNL_FCak6awSev3ueUK2DDa4guJ-zWzp. Date accessed: 04/30/2025.
51. Pillai, O. M., & Sundaramoorthy, S. (2022). Development of photocatalytic self-cleaning textiles using tin oxide and titanium dioxide nanoparticles mixture. Journal of the Textile Institute, 114(4), 674–681. https://doi.org/10.1080/00405000.2022.2062859
52. Pourzahedi, L. (2016). Environmental impact assessment of nanoparticles and Nano-enabled products using LCA frameworks. https://repository.library.northeastern.edu/files/neu:cj82n8838.
53. Radhakrishnan, S. (2021). Circular economy in textiles and fashion (pp. 163–202). Springer. https://doi.org/10.1007/978-981-16-3698-1_6
54. Ragab, A.-R., Al-Gizawy, A., Al-Minyawi, O., Hassabo, A., & Mahmoud, M. (2024). Environmental impact of clothing manufacturing and the fashion industry. Egyptian Journal of Textile and Polymer Science and Technology, 0(0), 0. https://doi.org/10.21608/jtcps.2024.259253.1298
55. Ramírez-Escamilla, H. G., Rodríguez, M., Padilla-Rivera, A., Domínguez-Solís, D., & Campos-Villegas, L. E. (2024). Advancing toward sustainability: A systematic review of circular economy strategies in the textile industry. Recycling, 9(5), 95. https://doi.org/10.3390/recycling9050095
56. Rayhan, M. G. S., Khan, M. K. H., Shoily, M. T., Rahman, H., Rahman, M. R., Akon, M. T., Hoque, M., Khan, M. R., Rifat, T. R., Tisha, F. A., Sumon, I., Fahim, A. W., Uddin, M. A., & Sayem, A. S. M. (2022). Conductive textiles for signal sensing and technical applications. Signals, 4(1), 1–39. https://doi.org/10.3390/signals4010001
57. RESYNTEX (2023). https://www.prospex-institute.org/resyntex. Date accessed: 04/30/2025.
58. Sacchet, S., Valentini, F., Rizzo, C., Pò, R., & Fambri, L. (2025). High density polyethylene with phase change materials for thermal energy management. Energy Materials, 5(4). https://doi.org/10.20517/energymater.2024.112

59. Saha, K., Dey, P. K., & Kumar, V. (2024). A comprehensive review of circular economy research in the textile and clothing industry. Journal of Cleaner Production. https://doi.org/10.1016/j.jclepro.2024.141252
60. Samanta, K. K., Basak, S., & Chattopadhyay, S. K. (2014). Environment-friendly textile processing using plasma and UV treatment (pp. 161–201). Springer Singapore. https://doi.org/10.1007/978-981-287-065-0_6
61. Shahabi, R., Dehghani, M., Moosavi, S. A. J., Shahabi, B., Poordakan, O., Sadeghi, M., Aryan, L., Ghasempoor, A., Aghanasiri, F., Mohseni, M., & Mehravi, B. (2022). The effect of nanoparticles on pulmonary fibrosis: A systematic review and meta-analysis of preclinical studies. Archives of Environmental & Occupational Health, 77(8), 684–694. https://doi.org/10.1080/19338244.2021.2001637
62. Shen, L., Worrell, E., & Patel, M. K. (2010). Environmental impact assessment of man-made cellulose fibres. Resources, Conservation and Recycling, 55(2), 260–274. https://doi.org/10.1016/j.resconrec.2010.10.001
63. Singh, N., & Sheikh, J. (2020). Self-cleaning textiles based on Superhydrophobic Nanocoatings (pp. 729–757). Jenny Stanford Publishing. https://doi.org/10.1201/9781003055815-24
64. Sobolieva, T., & Dobropas, A. (2024). Sustainable business transformation in textile industry: Challenges and good practices, 128–133. https://doi.org/10.24263/edsd-2024-6-16
65. Summer, M., Ashraf, R., Ali, S., Bach, H., Noor, S., Noor, Q., Riaz, S., & Khan, R. R. M. (2024). Inflammatory response of nanoparticles: Mechanisms, consequences, and strategies for mitigation. Chemosphere, 142826. https://doi.org/10.1016/j.chemosphere.2024.142826
66. Verma, A., Arif, R., Jadoun, S., & Jadoun, S. (2021). Impact of nanotechnology on sustainable textile material and its application (pp. 165–172). Woodhead Publishing. https://doi.org/10.1016/B978-0-323-85204-3.00030-0
67. Wagaw, T., & Babu, K. M. (2023). Textile waste recycling: A need for a stringent paradigm shift. AATCC Journal of Research, 10, 376–385. https://doi.org/10.1177/24723444231188342
68. Wang, J. N., Luo, X., Wu, T., & Chen, Y. (2014). High-strength carbon nanotube fibre-like ribbon with high ductility and high electrical conductivity. Nature Communications, 5(1), 3848. https://doi.org/10.1038/NCOMMS4848
69. Wang, R., Guan, C. C., Jin, X., Chai, J., Zhou, F., Wang, N., & Song, B. (2024). Sweat pore-inspired functional nanofiber fabrics for sweat management and simultaneous detection. Small. https://doi.org/10.1002/smll.202408465
70. Wo, Z., Xie, Y., Sun, H., Su, Y., Ma, H., & Zhang, X. (2024). Fabrics modified with ag nanowire/cellulose Nanofibril composites for high conductivity, electromagnetic interference shielding, ultraviolet protection, and thermal insulation. ACS Applied Nano Materials. https://doi.org/10.1021/acsanm.4c01957
71. Worn Again Technologies. (2022). https://wornagain.co.uk. Date accessed: 04/30/2025.
72. Xiong, Z., Cui, K., Chen, C., Guo, Z., Chen, X., & Chen, Y. (2024). The effects of silver particle size on bacterioplankton community and potential function in aquatic environment. Journal of Environmental Chemical Engineering. https://doi.org/10.1016/j.jece.2024.112012
73. Xu, L., Zhuang, W., Xu, B., & Cai, Z. (2011). Superhydrophobic cotton fabrics prepared by one-step water-based sol–gel coating. Journal of the Textile Institute, 103(3), 311–319. https://doi.org/10.1080/00405000.2011.569238
74. Xue, Y., Wang, Z., Zhang, Y., & Liu, W. (2020). A biocompatible and sustainable anti-ultraviolet functionalization of cotton fabric with Nanocellulose and chitosan nanocomposites. Fibers and Polymers, 21(11), 2521–2529. https://doi.org/10.1007/S12221-020-1339-X
75. Zhu, J., Ge, Y., & Zhang, X. (2022). Textile applications of nanofibers. In A. L. Andrady & S. A. Khan (Eds.), Applications of polymer nanfibers. Wiley. https://doi.org/10.1002/9781119267713.ch2
76. Živković, D., Balanović, L., Mitovski, A., Talijan, N., Štrbac, N., Sokić, M., Manasijević, D., Minić, D., & Ćosović, V. (2014). Nanomaterials environmental risks and recycling: Actual issues. Reciklaza i odrzivi razvoj, 7(1), 1–8. https://doi.org/10.5937/ROR1401001Z

Advancing Textiles with Nanotechnology: Innovations, Applications, and Sustainability Trends

B. Subathra

Abstract Sustainable Nanotextile technology represents a ground breaking advancement in the textile industry, merging nanotechnology with eco-friendly practices to create high-performance materials with minimal environmental impact. Nanotextile products are innovative materials enhanced with nanotechnology to deliver advanced functionalities beyond traditional textiles. The objectives of nanotechnology encompass a wide range of scientific, industrial, and societal goals, aimed at leveraging the unique properties of nanoscale materials and structures. The products are designed for various industries and offer improved performance, durability, and sustainability. This interdisciplinary approach leverages nanoscale engineering to enhance textile properties such as durability, water repellency, antibacterial activity, UV protection, and self-cleaning capabilities, all while promoting resource efficiency. This paper also analyzes the consolidation of nanotechnology in the textiles market to assess worldwide trends and patent coverage. Here are some notable nanotechnology textile products and their applications that are collectively taken for discussion, such as antimicrobial products for medical and surgical products; self-cleaning for laboratory uniforms; water-repellent for sports/home textile products; UV-resistant fabrics for sportswear; lightweight, and high-strength fabrics for military wear.

Keywords Finishing · Performance · Nanotextile · Special clothing · Sustainability

B. Subathra (✉)
Department of Apparel & Fashion Design, PSG College of Technology, Coimbatore, India
e-mail: bsa.afd@psgtech.ac.in

S. S. Muthu (ed.), *Sustainable Nano Textile Technology*, Textile Science and Clothing Technology, https://doi.org/10.1007/978-3-032-05902-4_5

1 Introduction to Nanotechnology in Textiles

Nanotechnology, the science of manipulating matter at the molecular and atomic levels (typically less than 100 nm), has revolutionized numerous industries, including textiles. Textiles have evolved beyond their traditional functions by integrating nanoscale materials and processes, offering enhanced performance, durability, and even smart capabilities. In textiles, nanotechnology enables the development of innovative features such as water repellence, stain resistance, UV protection, and antimicrobial properties without altering the fabric's natural texture or appearance. These advancements are achieved by incorporating nanoparticles, Nanocoatings, or nanofibers into the fabric structure, enhancing its functional properties at a molecular level [1]. The growing interest in nanotechnology for textiles is driven not only by the desire for functional versatility but also by the increasing demand for sustainable solutions. Nano-enhanced textiles reduce resource consumption by extending product lifespan and introducing environmental-friendly processes, aligning with global sustainability goals [2]. As the textile industry adopts this transformative technology, the possibilities for innovation and application seem limitless, from healthcare and fashion to military and sportswear.

Nanotechnology refers to the manipulation and engineering of materials at the nanometer scale (1–100 nm), where unique physical, chemical, and biological properties emerge due to the materials reduced dimensions. In textiles, nanotechnology involves incorporating nanoscale structures, such as nanoparticles, nanofibers, or nanocoatings, to imbue fabrics with advanced functionalities without compromising their comfort, weight, or appearance [3].

At this scale, materials exhibit unique properties like increased surface area, enhanced mechanical strength, and improved conductivity, making nanotechnology ideal for transforming textiles. For instance, nanotechnology enables fabrics to resist stains, repel water, provide UV protection, or possess antimicrobial properties. These enhancements are achieved by embedding nanomaterials such as silver nanoparticles for antimicrobial effects or titanium dioxide nanoparticles for UV resistance into fibers or coatings [4].

The application of nanotechnology in textiles represents a paradigm shift, enabling the development of multifunctional fabrics that go beyond traditional roles of clothing to serve in healthcare, sports, military, and environmental protection. This emerging field bridges material science and engineering, offering innovative solutions to meet modern consumer demands and sustainability goals.

Nanotechnology in textiles involves the clothing roles of nanoscale materials and structures into fabric to achieve advanced properties and functionalities. A nanometer is one-billionth of a meter, and at this scale, materials exhibit unique mechanical, optical, and chemical properties that differ significantly from their bulk counterparts. These properties, such as enhanced strength, increased surface area, and unique reactivity, make nanotechnology a powerful tool for textile innovation (Fig. 1).

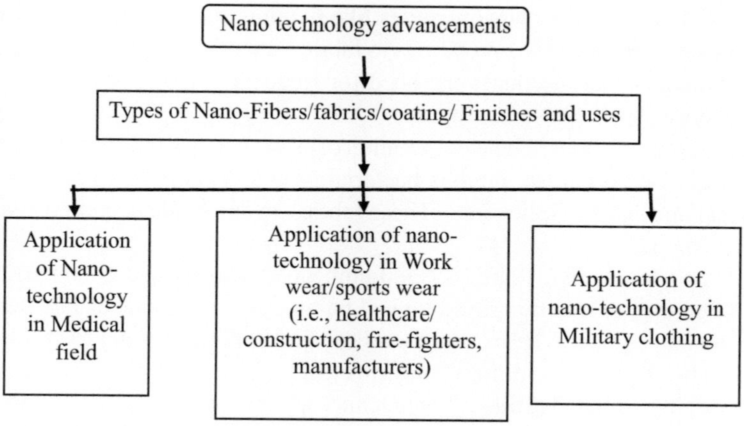

Fig. 1 Nano and textile innovation

In practice, nanotechnology is applied to textiles through several methods:

- Nanocoatings: Thin layers of nanoscale materials are applied to fabric surfaces to achieve effects like water repellency, stain resistance, and UV protection.
- Nanoparticles: Materials such as silver, titanium dioxide, and zinc oxide nanoparticles are embedded into fibers to provide antimicrobial, self-cleaning, or UV-blocking properties.
- Electro-spun Nanofibers: Fibers produced at the nanoscale are lightweight, durable, and breathable, offering improved performance for applications like filtration, sportswear, and protective gear [5].

The widespread adoption of nanotechnology in textiles has revolutionized traditional fabrics by providing enhanced functionality without compromising their natural aesthetics or comfort. For example, nanoscale silver particles are widely used in antimicrobial clothing, while nanocoatings ensure water resistance and extended durability in outdoor apparel. As the demand for high-performance, multifunctional, and sustainable textiles continues to grow, nanotechnology serves as a critical enabler of innovation, driving the textile industry toward smarter and more eco-friendly solutions.

2 Transforming Textiles through Nanoinnovations

Nanotechnology has transformed textiles from passive materials into dynamic, multifunctional products that meet the demands of modern industries. By leveraging the unique properties of nanoscale materials, traditional fabrics can be engineered to deliver enhanced performance, functionality, and sustainability. Nanoinnovations are revolutionizing textiles across various sectors, such as fashion, healthcare, sports, and defense, with ground breaking applications.

1. Self-Cleaning Textiles: Nanoparticles like titanium dioxide and silicon dioxide impart super-hydrophobic properties to fabrics, enabling them to repel water and dirt. This innovation reduces the need for frequent washing, conserving water and energy [4].
2. Antimicrobial Fabrics: The incorporation of silver nanoparticles, known for their antimicrobial properties, inhibits bacterial growth, making textiles suitable for medical and sports applications. These fabrics ensure hygiene and prevent odor, enhancing user comfort [3].
3. Smart and Interactive Textiles: Nano-embedded sensors and conductive fibers enable fabrics to monitor physiological parameters such as heart rate, temperature, and movement. These "smart textiles" find applications in healthcare wearables, fitness gear, and military uniforms [5].
4. Energy-Generating Textiles: Nanotechnology facilitates the development of energy-harvesting fabrics by integrating piezoelectric nanofibers or photovoltaic cells, enabling garments to generate power from movement or sunlight. This innovation is crucial for wearable electronics and off-grid energy solutions [6].
5. Enhanced Durability and Comfort: Nanocoatings provide textiles with increased resistance to wear and tear, UV radiation, and extreme weather conditions while maintaining breathability and softness. These features are particularly valuable in outdoor and performance wear.

By introducing nanoscale innovations, the textile industry is pushing the boundaries of what fabrics can achieve. These advancements not only enhance functionality but also support sustainable practices by reducing resource consumption and extending product lifespans.

3 Historical Evolution of Nano-Enhanced Textiles

The journey of nano-enhanced textiles can be traced back to the late twentieth century, when advancements in nanotechnology began to influence the materials science industry. The integration of nanotechnology into textiles has since evolved significantly, progressing through several key milestones [7].

3.1 Early Foundations (1980s–1990s)

The conceptualization of nanotechnology, first introduced by physicist Richard Feynman in 1959, laid the foundation for its application in materials science. By the 1980s, researchers began exploring the potential of nanoscale particles and fibers for enhancing the physical and chemical properties of materials. In textiles, initial experiments focused on incorporating nanomaterials like silica and titanium dioxide to improve fabric durability and UV resistance [11].

3.2 First Commercial Applications (2000s)

The early 2000s saw the commercialization of nano-enhanced textiles with practical applications. Products like stain-resistant and water-repellent fabrics emerged, leveraging nanocoatings made from fluorinated compounds and silicon-based nanoparticles. Nanoparticles of silver became widely used for their antimicrobial properties, making their way into sportswear, socks, and medical textiles [12].

3.3 Growth of Functional and Smart Textiles (2010s)

The integration of nanotechnology with electronics led to the development of smart textiles, capable of sensing and responding to environmental stimuli. Nanosensors and conductive nanomaterials enabled garments to monitor temperature, movement, and physiological parameters, revolutionizing applications in healthcare and fitness. Electrospinning techniques gained prominence during this period, allowing for the creation of nanofibers with enhanced mechanical strength and filtration capabilities.

3.4 Focus on Sustainability and Advanced Applications (2020s–present)

Recent advancements emphasize sustainable practices, including the use of biodegradable nanomaterials and eco-friendly manufacturing processes. "Nano-enhanced textiles now prioritize durability, recyclability, and reduced environmental impact" [14]. Emerging technologies include energy-harvesting fabrics and nanomaterials for environmental protection, such as pollution-absorbing textiles and water purification filters.

As nanotechnology continues to evolve, its integration with artificial intelligence, machine learning, and IoT is expected to further transform textiles. The focus is shifting toward creating multifunctional, adaptive, and sustainable solutions that redefine the role of fabrics in daily life.

4 Cutting-Edge Innovations in Nanotextiles

Nanotechnology has propelled textiles into a new era of functionality, enabling fabrics to deliver performance that was once unimaginable. Here are some of the most significant innovations in nanotextiles that are shaping industries and redefining fabric capabilities [8].

4.1 Super-Hydrophobic and Self-Cleaning Fabrics

Nanotechnology has made it possible to create fabrics with super-hydrophobic surfaces, mimicking the lotus leaf effect. Nanoparticles like titanium dioxide and silicon dioxide are applied as coatings, enabling textiles to repel water, dirt, and stains. "These self-cleaning fabrics significantly reduce the need for frequent washing, making them eco-friendly and ideal for outdoor and active wear" [14].

4.2 Antimicrobial and Odor-Resistant Textiles

The incorporation of nanoparticles such as silver, zinc oxide, and copper oxide has resulted in fabrics with strong antimicrobial properties. These textiles inhibit bacterial growth, ensuring hygiene and odor control, particularly in medical, sportswear, and everyday clothing. Silver nanoparticles, in particular, are widely used due to their effectiveness in combating bacteria and fungi.

4.3 Smart and Responsive Textiles

Nano-embedded sensors and conductive nanomaterials are driving the development of smart textiles that can monitor vital signs, temperature, and environmental conditions. For instance, nanofibers integrated with graphene enable energy storage and conductivity for wearable electronics. "Nanosensors in clothing monitor physiological parameters such as heart rate, paving the way for healthcare and fitness innovations" [12].

4.4 UV-Resistant Textiles

Nanoparticles like titanium dioxide and zinc oxide are used to create UV-blocking fabrics. These materials provide effective protection against harmful ultraviolet radiation while maintaining the breathability and comfort of the fabric, making them popular in outdoor and summer wear

4.5 Energy-Harvesting Fabrics

Advances in nanotechnology have enabled textiles to generate and store energy. Piezoelectric nanofibers embedded in fabrics harvest energy from body movement, while photovoltaic nanomaterials allow textiles to convert sunlight into usable energy. These energy-harvesting fabrics are critical for powering wearable devices and off-grid applications.

4.6 Nano-enhanced Filtration Materials

Nanofiber membranes and coatings are used to produce high-performance filtration textiles. These materials offer exceptional air and water filtration capabilities, capturing ultrafine particles and harmful contaminants. "Nano-enhanced filtration textiles are widely used in healthcare (e.g., face masks) and environmental applications" [14].

4.7 Lightweight and High-Strength Nanofibers

Electrospinning techniques are used to produce nanofibers with superior mechanical strength and lightweight properties. These fabrics find applications in aerospace, automotive, and protective clothing, where high strength-to-weight ratios are essential [12].

Nanocoatings have revolutionized the textile industry by imparting functional properties to fabrics without altering their inherent look or feel. These advanced coatings enhance fabric performance, making it more durable, easier to maintain, and protective against environmental factors. Nanotechnology applied to textiles can provide superior water-repellent, stain-resistant, and UV-protective properties, making these fabrics highly desirable across various industries.

4.8 Water-Repellent Nanocoating

Water-repellent nanocoatings are a key application of nanotechnology in textiles, aimed at improving fabric performance by preventing water penetration. Research in this area has focused on developing coatings that maintain fabric breathability while providing long-lasting water resistance. Below is a sample research overview on water-repellent nanocoatings for textiles [9].

5 Antibacterial and Antimicrobial Textiles in Medicine

Nanotechnology has significantly improved medical textiles by integrating antibacterial and antimicrobial properties. These textiles help prevent infections, reduce hospital-acquired diseases, and enhance patient care.

Antibacterial textiles are infused with nanoparticles such as:

- Silver (Ag) Nanoparticles—Disrupt bacterial cell membranes, preventing their growth.
- Copper (Cu) Nanoparticles—Effective against bacteria and fungi, offering long-lasting antimicrobial effects.
- Zinc Oxide (ZnO) Nanoparticles—Provide antibacterial and UV-protective properties.
- Graphene Oxide (GO) Nanoparticles—Offer broad-spectrum antimicrobial action and high durability.

5.1 Applications in the Medical Field

Silver nanoparticles are used as wound Dressings in the medical field. Silver-infused nanofiber dressings prevent bacterial infections and promote faster healing. Some advanced wound dressings release antimicrobial agents gradually for prolonged effectiveness. It also used as Hospital Bed Linens & Curtains with a special finish. Hospital environments are highly prone to bacterial and viral contamination, making infection control crucial. Nano-coated hospital bed linens and curtains provide antimicrobial, self-cleaning, and fluid-resistant properties, significantly reducing hospital-acquired infections (HAIs). Antimicrobial coatings on linens and curtains reduce the risk of cross-contamination in hospitals. Self-cleaning properties minimize the need for frequent washing, improving hygiene [10].

Nano-enhanced hospital textiles provide an added layer of protection against infections and contamination in healthcare settings. Patient bed linens and pillow covers are designed to prevent cross-infection in shared hospital beds, ensuring a hygienic environment for patients. Self-sanitizing privacy curtains help reduce the need for frequent washing by actively eliminating bacteria and viruses on their surface [11]. Surgical drapes and gowns play a crucial role in maintaining sterility in operating theaters, minimizing the risk of infections during medical procedures. Additionally, specialized emergency and ICU linens offer extra protection in critical care units, where infection control is essential for patient safety.

Surgical gowns and scrubs are being revolutionized with nanotechnology by incorporating nanoparticle coatings that prevent bacterial adhesion and enhance sterility. These advanced coatings create a protective barrier that minimizes the risk of contamination, ensuring a higher level of hygiene in medical environments. Additionally, these nano-treated fabrics are designed to maintain their effectiveness even after multiple washes, providing long-term protection and durability. This innovation not only improves infection control in hospitals but also enhances the safety and comfort of healthcare professionals [12].

Nanotechnology is transforming medical textiles by enhancing protection, durability, and hygiene in healthcare settings. Surgical gowns provide sterile protection in operation theaters, reducing the risk of contamination during procedures. Scrubs for healthcare workers, including doctors, nurses, and technicians, are designed for daily wear, offering comfort and antimicrobial properties to maintain cleanliness. Both disposable and reusable PPE benefit from nanotech enhancements, ensuring superior defense against pathogens while maintaining breathability and durability [13]. Additionally, specialized textiles used in ICU and high-risk areas provide enhanced infection control, safeguarding critical care workers and patients from cross-contamination in vulnerable environments.

The future of medical textiles is rapidly evolving with ground breaking innovations in nanotechnology. Smart gowns embedded with biosensors are being developed to detect bacterial contamination in real time, allowing for proactive infection control in healthcare settings. Eco-friendly nanocoatings are also gaining traction, offering biodegradable antimicrobial textiles that reduce environmental impact while maintaining high levels of protection [14]. Additionally, graphene-based PPE is emerging as a game-changer, providing lightweight yet ultra-protective nanofabrics designed to enhance safety and comfort for medical professionals. These advancements are set to revolutionize the healthcare industry, improving both patient care and the working conditions of frontline workers. Would you like details on specific brands, research studies, or manufacturing processes for nano-coated surgical gowns.

5.2 *Face Masks and PPE Kits*

Face masks and personal protective equipment (PPE) kits are essential for frontline healthcare workers and the general public. Nanocoating technology enhances their antimicrobial, antiviral, and filtration properties, offering superior protection against airborne pathogens, including bacteria, viruses, and pollutants. Nano-coated masks offer enhanced bacterial and viral protection.

Some masks include moisture-wicking and breathability features for comfort. Nanotechnology is revolutionizing healthcare and public safety by enhancing personal protective equipment (PPE) for frontline workers and the general public. N95 and surgical masks now incorporate nanoenhancements for superior filtration and antimicrobial protection, ensuring better defense against airborne pathogens. Reusable nanofiber masks provide everyday protection with long-lasting antimicrobial effects, reducing waste while maintaining effectiveness. Complete PPE kits, including gowns, gloves, and face shields, offer full-body protection in hospitals and laboratories, safeguarding healthcare professionals from hazardous exposure. Additionally, smart masks embedded with biosensors are being developed to monitor air quality and detect contamination levels in real time, further advancing safety measures in medical and public health environments [12].

The future of personal protective equipment (PPE) is being shaped by cutting-edge nanotechnology, enhancing both safety and sustainability. Self-sterilizing PPE, including masks and gowns, is being developed with specialized nanocoatings that actively disinfect surfaces, reducing the risk of contamination and extending usability. AI-integrated masks equipped with wearable technology can detect respiratory issues in real time, providing early warnings for potential health concerns. Additionally, eco-friendly nanofabrics are emerging as a sustainable solution, offering biodegradable PPE that minimizes environmental impact without compromising protection. These innovations are set to revolutionize public health and medical safety.

5.3 *Benefits of Antibacterial Nanotextiles*

Antibacterial nanotextiles are transforming healthcare by offering enhanced protection, durability, and hygiene. These advanced fabrics prevent bacterial infections and cross-contamination, significantly reducing the risk of hospital-acquired infections (HAIs). Their nanocoatings not only improve the longevity of medical textiles but also provide self-cleaning and odor-resistant properties, ensuring a cleaner and safer environment for both patients and healthcare workers. With these innovations, medical facilities can maintain higher hygiene standards while reducing the frequency of linen replacement and washing [15].

Nanocoating in wound dressings involves embedding or applying nanoscale materials to enhance their healing properties. These advanced coatings create an antimicrobial barrier, preventing bacterial infections and reducing the risk of complications. Bioactive nanoparticles play a crucial role in promoting tissue regeneration, accelerating the healing process, and minimizing scarring. Additionally, nanocoatings enable controlled drug release, ensuring a sustained therapeutic effect for optimal wound recovery. By maintaining a moist healing environment, these innovative dressings further support faster tissue repair while preventing excessive dryness or maceration.

5.4 *Types of Nanoparticles Used in Wound Dressings*

Silver Nanoparticles (AgNPs)

Silver nanoparticles (AgNPs) are widely used in wound care due to their powerful antimicrobial properties. They work by disrupting bacterial cell membranes, effectively killing microbes on contact and preventing infections. AgNPs are particularly valuable in combating antibiotic-resistant bacteria like MRSA, making them a crucial component in modern medical treatments. Their high efficacy and biocompatibility make them a common choice for burn wound dressings, where infection control and accelerated healing are essential [16].

Copper Nanoparticles (CuNPs)

Copper nanoparticles (CuNPs) play a dual role in wound healing by combining strong antimicrobial properties with regenerative benefits. They effectively inhibit bacterial growth while simultaneously promoting angiogenesis, the formation of new blood vessels, which is crucial for tissue repair. Additionally, CuNPs enhance collagen synthesis, strengthening the skin's structural integrity and accelerating the healing process. Due to these properties, they are commonly incorporated into dressings for chronic wounds, such as diabetic foot ulcers, where improved healing and infection prevention are critical.

Zinc Oxide Nanoparticles (ZnO NPs)

Zinc oxide nanoparticles (ZnO NPs) offer a multifunctional approach to wound healing with their antimicrobial, anti-inflammatory, and UV-protective properties. They help prevent infections while reducing inflammation, creating an optimal healing environment. ZnO NPs also stimulate wound closure by promoting cell proliferation and tissue regeneration, accelerating the recovery process. Due to these benefits, they are commonly used in the management of chronic wounds, where prolonged healing times and infection risks require advanced therapeutic solutions.

Graphene Oxide (GO) and Carbon Nanotubes (CNTs)

Graphene Oxide (GO) and Carbon Nanotubes (CNTs) have emerged as promising nanomaterials in advanced wound dressings due to their exceptional physicochemical properties. Their broad-spectrum antimicrobial effects help reduce the risk of infections, making them highly effective in promoting a sterile wound environment. Additionally, their remarkable durability and flexibility contribute to enhanced mechanical strength, ensuring that the dressing remains intact and adapts well to different wound sites. Furthermore, GO and CNTs facilitate oxygen diffusion, which plays a crucial role in cellular metabolism and tissue regeneration, ultimately accelerating the wound healing process. These properties make GO- and CNT-based wound dressings highly beneficial for improving patient outcomes in wound care management.

Chitosan-Based Nanoparticles

Chitosan, derived from natural biopolymers, is biodegradable. It exhibits strong antimicrobial properties and aids in blood clotting, making it useful in bioactive wound dressings [17].

5.5 Enhanced Antimicrobial Protection

Enhanced antimicrobial protection plays a critical role in modern wound management by directly targeting pathogens at the site of injury. This approach is designed to kill bacteria and fungi on contact, significantly reducing the microbial load and preventing the formation of biofilms that can complicate healing. By eliminating these pathogens immediately, the treatment minimizes the likelihood of infection and the subsequent need for systemic antibiotics. This reduction in systemic antibiotic use not only lowers the risk of adverse side effects but also helps combat the growing issue of antibiotic resistance, paving the way for more sustainable and effective wound care solutions.

Faster healing and tissue regeneration are essential for effective wound care, particularly in complex or slow-healing wounds. Advanced wound dressings incorporating innovative materials like Graphene Oxide (GO) and Carbon Nanotubes (CNTs) promote cell growth and stimulate collagen formation, both of which are crucial for tissue repair and regeneration. By accelerating these biological processes, these dressings help restore skin integrity more efficiently, making them ideal for treating chronic wounds, burns, and post-surgical wounds. Their ability to enhance the natural healing process not only improves patient outcomes but also reduces recovery time, minimizing the risk of complications and the need for prolonged medical intervention controlled drug delivery [18].

- Slow and sustained release of therapeutic agents.
- Reduces the frequency of dressing changes.

Nano-enhanced wound dressings play a crucial role in maintaining an optimal healing environment by balancing moisture retention and oxygenation. These advanced dressings help regulate hydration levels, preventing excessive dryness that can slow down recovery while also avoiding maceration, which can weaken surrounding tissue. By ensuring proper oxygen diffusion to the wound site, they promote faster cell regeneration and more efficient healing, ultimately improving patient outcomes.

Nano-enhanced wound dressings offer significant benefits in reducing scarring and providing pain relief. These advanced dressings help minimize inflammation and accelerate new tissue growth, promoting faster and more efficient healing. Additionally, some nanodressings are infused with analgesic nanoparticles that provide targeted pain relief, reducing discomfort for patients. By combining enhanced healing properties with pain management, these innovative dressings improve patient outcomes and overall recovery experiences [19].

6 Types of Nanotech Finishes on Workwear/Sportswear

Nanotechnology-based finishes on workwear enhance durability, functionality, and comfort while providing protection against bacteria, liquids, and environmental hazards. These finishes are widely used in industries such as healthcare, construction, firefighting, and manufacturing.

6.1 Antimicrobial and Antibacterial Finishes

Antimicrobial and antibacterial finishes are designed to prevent bacterial and fungal growth, effectively reducing odor and minimizing the risk of infections. These finishes utilize advanced technologies, including silver nanoparticles, copper nanoparticles, and zinc oxide, which are known for their potent antimicrobial properties. Such treatments are widely applied in various industries, including healthcare, where they enhance the hygiene of medical scrubs, lab coats, and personal protective equipment (PPE) used in hospitals. They are also essential in workwear for food processing facilities and cleanrooms, where maintaining a sterile environment is crucial. Additionally, antimicrobial finishes are incorporated into sportswear and military uniforms to prevent microbial buildup caused by sweat and prolonged use. An example of this technology in action is SteriTouch, a fabric coating infused with silver nanoparticles, which provides long-lasting antimicrobial protection for textiles and garments.

6.2 Water-Repellent and Stain-Resistant Finishes

Water-repellent and stain-resistant finishes are designed to create a hydrophobic surface that effectively repels water, oil, and stains, enhancing the durability and cleanliness of fabrics. These finishes leverage advanced technologies such as silica nanoparticles, fluoropolymers, and titanium dioxide (TiO_2) to provide long-lasting protection against liquid penetration and contamination. They are widely used in outdoor and industrial workwear, including garments for construction workers and oil rig personnel, where exposure to harsh environments is common. In healthcare settings, self-cleaning hospital uniforms and patient gowns benefit from these treatments, helping to maintain hygiene and reduce the risk of cross-contamination [16]. Additionally, laboratory coats and food industry apparel incorporate these finishes to ensure cleanliness and longevity in high-exposure environments. An example of this technology is Nano-Tex Resists Spills™, a water-repellent fabric treatment that enhances clothing performance by preventing liquid absorption and staining.

6.3 Fire-Resistant (FR) Nanocoatings

Fire-resistant (FR) nanocoatings are designed to enhance fabric safety by providing flame resistance and reducing flammability. Utilizing advanced technologies such as nanoclay and phosphorus-based nanocoatings, these treatments create a protective barrier that slows down the spread of flames and minimizes heat damage. They are essential in high-risk environments, particularly in firefighter gear and military uniforms, where exposure to extreme heat and fire hazards is a constant threat. Additionally, FR nanocoatings are widely used in industrial safety suits for professionals engaged in welding, electrical work, and other hazardous occupations. An example of this technology is NanoFR™, a specialized coating that enhances the fire-resistant properties of textiles, offering greater protection and safety for workers in demanding environments [17].

6.4 UV Protection Finishes

UV protection finishes are designed to block harmful ultraviolet (UV) radiation, helping to reduce skin damage and prevent fabric degradation caused by prolonged sun exposure. These finishes utilize advanced technologies such as titanium dioxide (TiO_2) and zinc oxide nanoparticles, which effectively absorb and scatter UV rays, enhancing the durability and safety of textiles. They are particularly beneficial in outdoor workwear for industries like construction and agriculture, where workers are frequently exposed to the sun. Additionally, protective uniforms for security and military personnel incorporate UV-resistant coatings to provide long-lasting defense against sun-related wear and tear. An example of this innovation is UV-Pro Tech Nanocoating, a specialized treatment applied to outdoor clothing to enhance UV protection and extend fabric lifespan.

6.5 Smart and Conductive Nanofinishes

Nanotech finished workwear offers a range of benefits that enhance both protection and performance in demanding environments. These advanced fabrics provide enhanced protection by guarding against bacteria, stains, and environmental hazards, ensuring long-lasting cleanliness and hygiene. Additionally, nanocoatings contribute to increased comfort by making the fabric breathable, lightweight, and moisture-wicking, keeping wearers dry and comfortable throughout the day. Durability is another key advantage, as nanotechnology strengthens the fabric, improving its longevity and overall performance. Furthermore, some eco-friendly nanocoatings help reduce the need for frequent washing, conserving water and energy while maintaining the workwear's effectiveness. Nano-Textiles in Military Applications Nanotechnology has significantly enhanced military textiles by providing lighter, stronger, and more functional fabrics for soldiers. These nanotextiles improve protection, durability, and adaptability in combat and extreme environments [18].

7 Applications of Nanotextiles in the Military

7.1 Bulletproof and Blast-Resistant Fabrics

Bulletproof and blast-resistant fabrics are designed to provide enhanced protection against bullets, shrapnel, and explosions while significantly reducing the weight of traditional body armor. These advanced materials incorporate cutting-edge nanotechnology to improve strength, flexibility, and impact resistance. Carbon nanotubes (CNTs) are used to create ultra-lightweight armor with exceptional tensile strength, while graphene-coated fabrics further enhance impact absorption and durability. Additionally, shear-thickening fluids (STFs), a unique liquid nanomaterial, remain flexible under normal conditions but instantly harden upon impact, providing adaptive protection [19]. These technologies are applied in next-generation protective gear, such as Dragon Skin Armor, which integrates nano-reinforced ceramic plates for superior ballistic resistance, and Liquid Armor, an STF-based flexible body armor that offers enhanced mobility without compromising safety.

7.2 Chemical and Biological Warfare Protection

Chemical and biological warfare protection textiles are designed to create an advanced barrier against hazardous agents, including toxic gases, biohazards, and chemical weapons. These protective materials leverage nanotechnology to enhance safety while maintaining comfort and functionality. Nanoporous membranes play a crucial role by blocking harmful chemical toxins while allowing air circulation, ensuring breathability for the wearer. Additionally, nanosilver and zinc oxide coatings provide antimicrobial properties, preventing the spread of biological contaminants [20]. These innovations are widely applied in military and emergency response gear, such as nano-coated Chemical, Biological, Radiological, Nuclear (CBRN) suits, which offer enhanced protection in high-risk environments. Furthermore, gas masks equipped with nanofilters significantly improve toxin filtration, ensuring superior defense against airborne threats.

7.3 Adaptive Camouflage and Stealth Clothing

Adaptive camouflage technology is designed to change color or texture to match the surrounding environment, significantly reducing visibility to enemies and enhancing stealth capabilities. This innovation utilizes advanced nanotechnology, including metamaterials and nanophotonic coatings, which can bend light to create near-invisibility effects [21]. Additionally, graphene-based dynamic camouflage enables real-time color adjustments by responding to environmental changes, allowing for seamless blending into various terrains. These breakthroughs are applied in military gear, such as Adaptive Nano-Camo Uniforms, which shift colors dynamically to match their surroundings, providing superior concealment. Furthermore, specialized nanocoatings for infrared suppression help reduce heat signature detection, making it more difficult for thermal imaging systems to track personnel and equipment in combat zones.

7.4 Self-Cleaning and Water-Repellent Uniforms

Self-cleaning and water-repellent uniforms are designed to maintain cleanliness and hygiene by preventing the build-up of dirt, bacteria, and moisture, especially in extreme conditions. These advanced textiles utilize super-hydrophobic nanocoatings that repel water, mud, and oil, ensuring that uniforms remain dry and stain-free. Additionally, titanium dioxide (TiO_2) nanoparticles play a crucial role in breaking down dirt and bacteria when exposed to sunlight, providing a continuous self-cleaning effect. This technology is widely applied in various performance-enhancing garments, such as lotus-effect nanofabric uniforms that effectively repel contaminants, keeping wearers clean and comfortable. Another key innovation is odor-resistant nanosocks, which incorporate antibacterial nanocoatings to reduce bacterial growth and prevent unpleasant odors, making them ideal for prolonged use in harsh environments [22].

7.5 Smart and Wearable Sensor-Embedded Textiles

Smart and wearable sensor-embedded textiles are revolutionizing military gear by integrating real-time health monitoring and battlefield awareness directly into soldiers' uniforms. These advanced fabrics utilize graphene-based biosensors to detect critical health parameters such as heart rate, hydration levels, and stress, allowing for early intervention in case of injury or fatigue. Additionally, conductive nanofibers enable seamless real-time communication through the fabric, reducing the need for bulky external devices. These innovations are applied in smart uniforms with embedded biosensors, which track soldier vitals and environmental hazards, enhancing safety and operational efficiency [23]. Furthermore, embedded radio communication textiles facilitate hands-free communication, ensuring seamless coordination on the battlefield without compromising mobility or awareness.

7.6 Fire-Resistant and Thermal Regulation Clothing

Advanced thermal and fire-resistant textiles are designed to protect soldiers from extreme temperatures, explosions, and burns, ensuring safety and comfort in harsh environments. These high-performance fabrics incorporate nanoclay and phosphorous coatings to enhance fire resistance, reducing the risk of injury in combat or hazardous situations. Additionally, phase-change materials (PCMs) play a crucial role in regulating body temperature by absorbing and releasing heat, providing adaptive thermal control in both hot and cold climates. These innovations are utilized in nano-FR combat gear, which offers fire-resistant protection for frontline soldiers, as well as PCM-based thermal uniforms that adjust to environmental conditions, making them ideal for deployment in desert or arctic regions [24].

Nanotextiles are revolutionizing modern military gear by enhancing soldier safety, performance, and adaptability in challenging environments. These advanced fabrics are lightweight yet highly durable, reducing fatigue while maintaining superior protection. Enhanced stealth capabilities, including dynamic camouflage and infrared suppression, allow soldiers to blend into their surroundings and evade detection. Improved survivability is achieved through cutting-edge materials that shield against bullets, chemical agents, and extreme temperatures. Additionally, integrated battlefield awareness is made possible with wearable sensors that provide real-time health and environmental data, offering tactical advantages. Maintenance is also significantly reduced, as self-cleaning and antimicrobial properties minimize the need for frequent washing. As a result, nanotextiles are making soldiers safer, stronger, and more adaptive to modern warfare.

7.7 Adaptive Camouflage and Stealth Clothing in Military Applications

Adaptive camouflage and stealth clothing use nanotechnology to dynamically change color, texture, and thermal signatures, making soldiers and vehicles nearly invisible to enemies. These innovations are inspired by natural camouflage mechanisms, such as the color-changing ability of chameleons and cephalopods (octopuses, squids).

8 Technologies in Adaptive Camouflage and Stealth Clothing

8.1 Nanophotonic and Metamaterial Coatings

Nanophotonic and metamaterial coatings are cutting-edge technologies designed to manipulate light waves, bending them around objects to create invisibility or seamless environmental blending. These coatings utilize metamaterials, which are artificially engineered to control electromagnetic waves, allowing for unprecedented control over light behavior. Additionally, graphene and carbon nanotubes (CNTs) play a crucial role by absorbing a broad spectrum of light wavelengths, including infrared (IR), making them highly effective in evading night-vision detection. One notable example of this technology is "Quantum Stealth" by Hyper-Stealth Biotechnology, a revolutionary light-bending material that creates the illusion of invisibility by redirecting light around objects [25]. This breakthrough in stealth technology has significant military and tactical applications, enhancing concealment and reducing detection risks in combat environments.

8.2 Thermochromatic and Electrochromic Nanofibers

Adaptive nanomaterials are revolutionizing military textiles by enabling fabrics to change color and temperature in response to electrical signals or environmental conditions. These materials utilize liquid crystal nanofilms, which shift colors based on temperature changes, and electrochromic polymers, which alter their hue when voltage is applied. This technology allows soldiers to dynamically adjust their camouflage patterns to match their surroundings, enhancing stealth and operational effectiveness [26]. A notable example is the U.S. Army's Future Soldier Initiative, which is developing advanced uniforms capable of real-time camouflage adaptation, providing soldiers with enhanced concealment and environmental adaptability on the battlefield.

8.3 Infrared Suppression Nanocoatings

Infrared suppression nanocoatings are advanced materials designed to absorb and scatter infrared (IR) radiation, making it difficult for soldiers to be detected by night vision or thermal imaging systems. These coatings utilize titanium dioxide (TiO_2) and carbon-based nanomaterials, which effectively absorb infrared radiation, reducing visibility to heat-sensing devices. Additionally, aerogel-based coatings play a crucial role in minimizing heat signatures by insulating against body heat or engine emissions, further enhancing stealth capabilities. A notable example of this technology is "Metaflex Coatings," a nanomaterial developed in the UK that effectively blocks thermal signatures, providing soldiers and military equipment with an added layer of protection against infrared detection on the battlefield.

8.4 Biomimetic Camouflage (Inspired by Nature)

Biomimetic camouflage, inspired by nature, mimics the adaptive skin of cephalopods such as squid, octopuses, and cuttlefish, which can instantly change color, texture, and brightness to blend seamlessly into their surroundings. This advanced technology utilizes bio-engineered chromatophores, which are nanoscale pigment cells capable of dynamically shifting colors in response to environmental stimuli. Additionally, elastomer-based flexible nanomaterials enable fabrics to adjust their texture, further enhancing concealment by matching the surrounding terrain. A ground breaking example of this innovation is the Cephalopod-Inspired Stealth Suit developed by MIT researchers, which incorporates synthetic chromatophores to replicate the rapid and highly effective camouflage abilities of octopuses, offering unprecedented stealth advantages for military and tactical applications.

8.5 Smart Textiles with Embedded Sensors

Smart textiles with embedded sensors are revolutionizing military camouflage by integrating nanosensors directly into fabrics to detect environmental changes and dynamically adjust camouflage in real time. These advanced textiles utilize graphene-based nanocircuits, which enable instant control over camouflage patterns, allowing soldiers to blend seamlessly into their surroundings. Additionally, piezoelectric nanofibers play a crucial role by converting mechanical signals, such as movement and temperature fluctuations, into adaptive camouflage responses. A notable example of this cutting-edge technology is the Adaptive Battlefield Uniform, a DARPA project designed to incorporate smart textile systems with embedded sensors, providing soldiers with enhanced stealth and survivability on the battlefield [27].

8.6 Real-World Military Applications and Projects

Quantum Stealth, developed by HyperStealth Biotechnology, is a revolutionary light-bending material that renders objects nearly invisible without requiring any power source. This advanced technology works by manipulating light waves to conceal objects, making it ideal for stealth applications. It has been explored for use in special forces cloaking prototypes, providing soldiers with enhanced concealment in combat environments [28]. Currently, Quantum Stealth has been successfully demonstrated and is undergoing military testing to assess its effectiveness for real-world tactical operations.

8.7 U.S. Army's Future Soldier Initiative

The U.S. Army's Future Soldier Initiative is focused on developing advanced color-changing uniforms that can instantly switch camouflage patterns to adapt to different environments. This cutting-edge technology utilizes liquid crystal nanotechnology, enabling soldiers to seamlessly blend into urban, jungle, and other combat settings with minimal effort [5]. By providing real-time adaptability, these uniforms are expected to significantly enhance stealth and operational effectiveness, giving soldiers a tactical advantage in diverse and rapidly changing battlefield conditions [29].

9 Conclusion

Nanotechnology is revolutionizing the textile industry by enhancing fabric properties, enabling smart functionalities, and promoting sustainability. Innovations such as self-cleaning fabrics, moisture-wicking materials, and antimicrobial textiles are reshaping fashion, healthcare, and industrial applications. Additionally, nanomaterials contribute to eco-friendly practices by reducing water and energy consumption, extending textile lifespan, and enabling biodegradable alternatives.

As research advances, the integration of nanotechnology in textiles will continue to expand, addressing challenges related to safety, scalability, and environmental impact. A collaborative approach involving scientists, manufacturers, and policymakers is essential to ensure responsible innovation that balances performance with sustainability. The future of textiles lies in harnessing nanotechnology to create smarter, more durable, and environmentally conscious fabrics, paving the way for a more advanced and sustainable textile industry. The speciality wear are easy to get spoiled if used as military uniform, lab coats, sportswear or work wear, so application of special finishes on the surface will be more helpful to avoid stain/damage on material and support to extend its life.

The application of nanofibers in textiles significantly enhances sustainability by enabling the development of high-performance fabrics with minimal environmental impact. Nanofibers, due to their high surface area-to-volume ratio and lightweight nature, can be engineered to provide properties such as water repellency, UV protection, antibacterial effects, and improved strength without relying on heavy chemical treatments or additional layers of material. This reduces the need for resource-intensive processes and prolongs the lifespan of textiles, lowering overall consumption and waste. Additionally, biodegradable and bio-based nanofibers offer eco-friendly alternatives to synthetic fibers, further supporting circularity in the textile industry.

References

1. Deepa Suhag. Handbook of Biomaterials for Medical Applications: Applications (2). 2024 2024. ISBN-13: 978-9819759057 ISBN-10. doi:https://doi.org/10.1007/978-981-97-5906-4.
2. Madhu, C. R. (2024). *Nano technology in textile finishing-Advancements and applications*. https://doi.org/10.1007/978-981-97-2696-7
3. Singh, N., Madhu, A., & Yadav, M. (2024). *Nano-finishing in apparel textiles, materials horizons: From nature to nanomaterials nanotechnology in textile finishing* (pp. 415–440).
4. Ibrahim, M., Mondal, H., et al. (2022). *Metal and metal oxides nanoparticles in healthcare and medical textiles, Medical Textiles from Natural Resources* (The textile institute book series) (pp. 341–371). https://doi.org/10.1016/B978-0-323-90479-7.00010-5
5. Zille, A., Almeida, L., Amorim, T., Carneiro, N., Esteves, M. F., Silva, C. J., & Souto, A. P. (2014). Application of nanotechnology in antimicrobial finishing of biomedical textiles. *Materials Research Express, 1*(3), 032003. https://doi.org/10.1088/2053-1591/1/3/032003. End original source.
6. Khan, M. S. H., & Rahaman, M. T. (2025). *Eco-friendly organic nanomaterials for multifunctional textiles: Sources, applications, recent advances and future prospects towards sustainability* (Vol. 22).
7. Uddin, F. (2008). Some aspects in advancement of textile nano finishing. In *86th Textile-Institute World Conf., Conf. Proceed. 2, U291–U307*. Polytechnic Univ., Inst. Textiles and Clothing Press.
8. Anik, H. R., Tushar, S. I., Mahmud, S., Khadem, A. H., Sen, P., & Akter, M. (2024). Into the revolution of nanofusion: Merging high performance and aesthetics by nanomaterials in textile finishes. *Advanced Materials Interfaces, 12*, 2400368. https://doi.org/10.1002/admi.202400368
9. Yingngam, B. (2024). Nano-coatings for cosmeto textiles. *Sustainable Approach to Protective Nano-coatings*. https://doi.org/10.4018/979-8-3693-3136-1.ch006
10. Jinlian, H., Kumar, B., & Jing, L. (2020). *Handbook of fibrous materials*. , Print ISBN: 9783527342204| Online ISBN:9783527342587. Wiley-VCH Verlag GmbH & Co. KGaA. https://doi.org/10.1002/9783527342587
11. Bhushan, B. (2017). *Introduction to nanotechnology*. Springer.
12. Kaushik, A., & Singhal, P. (2020). Nanotechnology applications in textiles for sustainable development. *Journal of Textile Science and Technology, 5*(3), 101–112.
13. Rai, M., Yadav, A., & Gade, A. (2020). Current trends in nanotechnology for textile applications. *Advanced Functional Materials, 12*(4), 122–135.
14. Shahid-ul-Islam, & Butola, B. S. (2019). Nanotechnology for functional textiles: Applications and prospects. *Journal of Materials Science, 54*(15), 11309–11338.
15. Singh, N., Madhu, A., & Yadav, M. (2024). Nano-finishing in apparel textiles. In *Nanotechnology in textile finishing: Advancements and applications* (pp. 415–440). Springer.
16. Singh, S., Kamble, Z., & Neje, G. (2024). Types of nanomaterials used in textile finishing. In *Nanotechnology in textile finishing: Advancements and applications* (pp. 23–48). Springer.
17. Wang, R. (2007). *Multi-functional protective nano-finishing of textile materials*.
18. Anik, H. R., Tushar, S. I., Mahmud, S., Khadem, A. H., Sen, P., & Akter, M. (2025). Into the revolution of nanofusion: Merging high performance and aesthetics by nanomaterials in textile finishes. *Advanced Materials Interfaces, 12*(1), 2400368.
19. Gokarneshan, N., Gopalakrishnan, P. P., & Jeyanthi, B. Influence of nano finishes on the functional properties of textile materials. *International Journal of Basic and Applied Chemical Sciences*. ISSN 2277-2073.
20. Uddin, F. (2008). Some aspects in advancement of textile nanofinishing. In *86th textile-institute world Conf* (pp. U291–U307). Polytechnic Univercity, Institute of Textiles and Clothing Press.
21. Gulrajani, M. L., & Deepti, G. (2011). *Emerging techniques for functional finishing of textiles*.
22. Bashari, A., Shakeri, M., Shirvan, A. R., & Najafabadi, S. A. N. (2018). Functional finishing of textiles via nanomaterials. *Nanomaterials in the wet processing of textiles*, 1–70.

23. Gokarneshan, N., & Velumani, K. (2018). Significant trends in nano finishes for improvement of functional properties of fabrics. In *Handbook of renewable materials for coloration and finishing* (pp. 387–434). Scrivener Publishing LLC.
24. Vhanbatte, S. B., Landage, S. M., Wasif, A. I., Dansena, B., & Karche, N. V. (2017). Nanotechnology for antimicrobial finishing of textiles. *International Journal of Advanced Research in Engineering and Applied Sciences, 6*(6), 14–23.
25. Montazer, M., & Harifi, T. (2018). *Nanofinishing of textile materials*. Woodhead Publishing.
26. Madhu, A., Singh, N., & Yadav, M. (2024). Challenges and opportunities in nanotechnology for textile finishing. In *Nanotechnology in textile finishing: Advancements and applications* (pp. 487–511).
27. Mahmud, R., & Nabi, F. (2017). Application of nanotechnology in the field of textile. *IOSR Journal of Polymer and Textile Engineering, 4*(1), 01.
28. Mahmoud, Z. H., & Kianfar, E. (2024). Application of nano technology in the self-cleaning finishing of textiles: A review. *Journal of Textile Engineering and Fashion Technology, 6*(1), 01–13.
29. Ghazal, H., Khaleed, N., & Abdelaziz, E. (2023). Significance advantages, and disadvantages of nanotechnology in textile finishing. *Egyptian Journal of Chemistry, 66*(12), 467–482.

Sustainability and Functional Performance of Nanoscience in Textiles

S. Grace Annapoorani and K. Saranya

Abstract The use of nanotechnology in the apparel industry is not a recent development. Many apparel firms began adding silver nanoparticles to their goods in the middle of the 2000s. Because silver nanoparticles are antibacterial, they eliminate the microorganisms that produce unpleasant odours. The finished clothing requires fewer washings as these nanoparticles are added to the fabric to stop odour. These nano-infused textiles are still in demand today and come in various techniques, from t-shirts to socks. The field of nanofabrics is producing far more advanced technologies than only antibacterial nanoparticles. Self-cleaning textiles, water-repellent textiles, and clothing that may lessen odours by chemically altering chemical compounds that create unpleasant odours are further desirable clothing qualities that might be accomplished with nanotechnology. These developments would benefit from the unique characteristics of nanomaterials, especially their high surface area-to-volume ratio, which increases the exposure of active surfaces to the environment. It is possible that more recent nanofabric technologies have significant issues of their own. However, it is worthwhile to investigate the possible advantages of using textiles strengthened with nanotechnology. And this area of nanotechnology can only expand as long as science keeps improving and enabling us to solve environmental issues. However, several textile materials up to the nanometre scale have been successfully produced using nanotechnology and nanoscience, having extremely porous fibres and large surface area that demonstrate their viability for advanced applications. Because of their unique structures and nanoscopic size, these materials have exceptional capabilities that provide distinctive electrical, magnetic, optical, biological, thermal, and mechanical features. The nanofibres can be created using various methods, such as phase separation, centrifugal spinning, electrospinning, melt blowing, drawing, template synthesis, and biocomponent extrusion. The possible applications of nanofibres include tissue engineering, thermal-resistant materials, biomedical products, filtration down to the nanometre

S. G. Annapoorani (✉) · K. Saranya
Department of Textiles and Apparel Design, Bharathiar University, Coimbatore, Tamil Nadu, India
e-mail: gracetad@buc.edu.in

S. S. Muthu (ed.), *Sustainable Nano Textile Technology*, Textile Science and Clothing Technology, https://doi.org/10.1007/978-3-032-05902-4_6

scale, breathable sportswear, and protective apparel for military, chemical, and medical personnel. These cutting-edge application textiles may be produced effectively by using nanotechnology and reinforcing the textiles with nanofibre, sand, and nanoparticle additions. A notable development at present is the incorporation of environmentall-friendly nanomaterials, such biodegradable and bio-based nanoparticles, which promote more sustainable production methods. These innovations improve textile performance by providing qualities including water resistance, UV protection, and self-cleaning properties, in addition to addressing environmental concerns. It is anticipated that future developments will focus on improving the synthesis of nanomaterials, increasing production, and maintaining the sustainability of the entire lifespan. It is projected that new developments like smart features and circular economy strategies would further transform this industry.

Keywords Nanotechnology · Apparel industry · Environmentally friendly · Biodegradable · Sustainability and economy strategies

1 Introduction

The manipulation of materials at the size of individual atoms or molecules, including separation, consolidation, and deformation, is known as nanotechnology [1]. One of the best examples of an emerging technology is nanotechnology, which offers designed nanomaterials with enormous potential for producing goods with much improved performance [2]. Currently, nanomaterials are used in a wide range of commercial applications, such as energy storage devices, sports equipment, electronics, cosmetics, environmental cleanup, scratch-resistant paints, and surface coatings [3]. Nanofibres are used in many medical applications, such as tissue engineering, implant materials, wound dressings, drug delivery systems, and artificial organ components. Nanotextiles, fabrics engineered with nanoparticles to achieve enhanced functionalities like water repellency, antimicrobial properties, and UV protection, are increasingly being viewed through the lens of sustainability. The goal is to harness the innovative potential of nanotechnology while minimizing environmental impact and promoting ethical practices throughout the textile lifecycle.

A sustainable action to nanotextiles involves a holistic consideration of materials, processes, usage, and end-of-life management. This means moving beyond simply creating high-performance fabrics to ensuring that their production and disposal are environmentally sound and socially responsible [4].

2 Role of Nanomaterials in Textile Industry

Over the years, functional fabrics have garnered a lot of attention due to their remarkable qualities, such as flame resistance, UV protection, self-cleaning, and antibacterial activities. In this field, nanomaterials (NMs) have drawn increasing interest because of their special qualities and wide range of uses. NMs make it possible to create sophisticated functional textiles by modifying the characteristics of conventional fabrics [5]. Textile coatings and finishes now have a plethora of options thanks to nanotechnologies, which can provide improved functionality or completely new features. Improvements in chemical, biological, thermal, mechanical, optical, and electrical characteristics are among these breakthroughs. There are currently a number of textile goods on the market that are enabled by nanotechnology, including electrically conductive, self-cleaning, and antimicrobial materials [6]. The textile business has a lot of economic potential thanks to nanotechnology because traditional techniques for giving textiles new qualities frequently don't endure and may stop working after washing or wearing. However, because nanoparticles have a high surface area-to-volume ratio and higher surface energy, which enable them to connect with textiles more efficiently and retain functionality over time, nanotechnology guarantees improved longevity for fabrics [7]. Textile finishing uses nanoparticles, including metal oxides and ceramics, to alter surface characteristics and provide desired functionality. They are more effective than bigger particles in treatments because of their tiny size, which increases their surface area. Furthermore, nanoparticles are transparent and have no effect on the textile materials' colour or brightness [8]. The functionality and performance of textile materials have been profoundly altered by nanotechnology, which has also produced amazing advancements and drawn a lot of scientific interest. Both industrial and consumer applications make advantage of the dissemination of different nanomaterials. The market leader is nanosilver, which is followed by nano-Zn/ZnO and carbon nanotubes (CNTs) [9]. Nanomaterials can alter the physical characteristics of textiles when they are incorporated into them at the atomic and molecular level. Since conventional techniques for giving materials functionality sometimes lack durability and have effects that fade after washing or wearing, nanotechnology has enormous economic promise in the textile sector. On the other hand, nanotechnology uses nanoparticles with a high surface energy and a big surface area-to-volume ratio to guarantee long-lasting performance. Stronger adherence to textiles is made possible by these qualities, which leads to improved and long-lasting functioning [10]. The market for multipurpose smart and intelligent textiles is expanding quickly due to rising customer demand for high-performing, creative goods and technological improvements. This change is made possible by nanotechnology, which makes it possible to create fabrics with improved qualities and features that beyond conventional limits. Textiles may be made more durable and multipurpose by incorporating nanoscale elements into fibres and fabrics to increase their physical, chemical, and biological qualities. Medical and healthcare fabrics with antibacterial qualities and wound-healing capabilities, temperature-adaptive apparel that controls body temperature under changing conditions, and odour-resistant undergarments that contain antimicrobial nanoparticles like silver to stop the growth of bacteria that cause odours are a few examples of high-performance textiles created with nanotechnology. The potential uses of textiles in sectors including healthcare, sports, the military, and fashion have increased as a result of these advancements. The need for multifunctional smart textiles is anticipated to increase as this field of study develops, providing ground-breaking solutions to satisfy changing consumer demands [11]. In order to improve fabric qualities, the alteration and characterisation of textiles—especially via plasma and nanopretreatment—has attracted a lot of interest. Nanotechnology improves various fabrics, including those with UV protection, antistatic properties, sensors, antimicrobial properties, odour management, and water repellency. Nanomaterials have a greater potential to give textiles a variety of functions because of their high surface area-to-volume ratio and nanoscale size. (1) Carbon-based nanomaterials, such as graphene, carbon nanofibres, and carbon nanotubes; (2) Inorganic nanoparticles, such as metal oxides, metals, and nanoclays; (3) Core–shell nanoparticles; (4) Composite nanomaterials; (5) Hybrid nanomaterials; and (6) Polymeric nanomaterials are the most often utilized nanomaterials in textile applications. Advanced qualities including increased strength, durability, and multifunctionality are made possible by these materials in textiles [12].

Washing tests were used to determine the emissions of silver nanoparticles NM-300 K from textiles. The European Chemicals Agency's (ECHA) instructions on information requirements and chemical safety assessment served as the foundation for the environmental risk assessment that was conducted. No environmental threat of the silver nanoparticles NM-300 K released from textiles was observed, depending on the preconditions and scenarios chosen [13]. There are worries about the possible drawbacks of using nanoparticles and nanomaterials in smart fabrics. Toxic chemicals are frequently used in the production process, and when textiles are washed, nanoparticles may leak out and contaminate water supplies. Studies have shown, for instance, that silver nanoparticles from treated blankets may enter water systems, highlighting the dangers these materials pose to the environment [14]. Depending on their individual qualities and intended usage, green nanoparticles may be integrated into smart textiles in a variety of ways. Coating (e.g., dip-coating, spray-coating, or electrospinning), embedding, printing (e.g., inkjet, screen, or gravure printing), and weaving are common integration techniques [15]. The environment, sustainability, human health, and safety are all under danger from nanotextiles. Where nanoparticles are incorporated into textiles determines how exposed they are to outside stimuli, making them either more or less susceptible to such impacts [16].

3 Environmental Impact of Using nanomaterials in Textiles

In order to improve current product features and introduce new ones, research institutes and industry are now conducting intensive studies on engineered nanomaterials. Up until 2018, little was known about the possible effects of nanomaterials on the environment and human health, despite tremendous advances in the field. After being discharged into the environment, nanomaterials can be hard to find, which raises questions about their potential to create a number of ecological problems, particularly if cleanup techniques are dangerous or insufficient. To better understand their effects and guarantee safe handling, it is imperative to gather thorough data on the emissions and ambient concentrations of nanomaterials employed in nanotextiles. In order to comprehend how nanomaterials interact with biological systems and the environment, further study is required to examine the structure–function interactions of these materials, concentrating on their basic chemical characteristics, such as toxicity and functionality. For nanoparticles that provide possible exposure concerns during their manufacture or usage, thorough risk evaluations should also be carried out. By identifying risks, estimating the likelihood of exposure, and establishing safe usage procedures, these assessments will help realize the advantages of nanotechnology while protecting the environment and public health [17]. With qualities like antistatic, UV protection, flame retardancy, antibacterial effects, and self-cleaning capabilities, nanomaterials hold enormous promise for improving textiles. Studies indicate that they are appropriate for use in advanced textiles, notwithstanding difficulties in their commercial implementation. However, there are still potential and problems in the rapidly evolving field of nanotechnology. There are insufficient safety restrictions, especially with regard to toxicological concerns, as textile exposure to nanoparticles rises. Particularly for susceptible populations like children, improvements in monitoring methods are aiding in the evaluation of possible health hazards. In conclusion, even if nanoparticles have many advantages, their hazards must be recognized in order to guarantee their safe application in textiles [18].

4 Health Hazards of Using Nanomaterials in Textile Industry

Because exposure to artificial nanomaterials has been linked to a number of health hazards, such as cancer, genetic damage, lung inflammation, and circulatory abnormalities, nanosafety has grown in importance [19]. Ingestion, inhalation, and skin absorption are some of the ways that people might be exposed to engineered nanoparticles from textiles [20]. Because they are exposed to large amounts of designed nanoparticles over an extended period of time, workers in the textile industry are most at risk. The skin and respiratory system are the main pathways by which nanoparticles are absorbed. Once within the body, nanomaterials have a tendency to gather in organs including the liver, kidneys, spleen, and bones. The fictitious model of engineered nanoparticles (ENPs) shows how they travel through the human body, emphasizing the poisonous and detrimental effects on the skin, nose, and throat [21]. Because of its enormous potential and economic significance, nanotechnology has emerged as a top research priority in many nations. However, the environment and human health may be at danger due to the distinctive features of some nanomaterials, such as differences in size, shape, and chemical makeup. The four main areas of textile applications—nanofibres, nanocomposites, nanofinishing, and nanocoatings—are examined. The presence of carcinogenic and hazardous compounds in textile materials must be carefully assessed since they frequently come into direct and extended contact with human skin. For nanoparticles to evolve safely and sustainably in the future, it is crucial to comprehend both their possible advantages and unforeseen hazards [22]. Because textiles come into continual touch with us throughout the day, they are sometimes referred to as our second skin. They provide various practical and defensive functions in addition to being aesthetically pleasing and beautiful. Investigating the possible presence and consequences of hazardous and carcinogenic chemicals in textiles is crucial because of their extended and direct contact with human skin. To assure safety, special consideration and assessment are needed for engineered nanomaterials, a contemporary and potentially hazardous class of compounds [23]. Nanomaterials have enormous potential to raise living standards and spur European industry. But they can provide health and safety issues as well as possible environmental hazards. Although there are still many scientific questions about the safety of nanomaterials, the Scientific Committee on Emerging and Newly Identified Health Risks states that they are not intrinsically harmful. This calls for comprehensive safety analyses conducted on an individual basis. The health hazards connected to "nanotextiles"in the textile industry are still not well understood, which emphasizes the need for further targeted research in this field. Concerns regarding possible negative effects are raised by the usage of nanoparticles and nanomaterials in the creation of smart textiles. Their production frequently involves toxic chemicals, and when the final items are washed, nanoparticles may leak out and contaminate water supplies. Research has demonstrated that silver nanoparticles from treated blankets, for instance, can wash away and penetrate aquatic ecosystems, underscoring the dangers such materials pose to the environment [24].

5 The Promise of Nanoscale Solutions

Sustainability is a critical global concern, demanding innovative approaches across industries. Finishing processes, which modify the surface properties of materials, often involve resource-intensive and environmentally impactful methods. Nanotechnology, by manipulating matter at the atomic and molecular level, offers transformative solutions for creating sustainable finishes. This field holds immense potential to revolutionize how we treat materials, leading to reduced resource consumption, minimized waste, and enhanced product performance. This exploration will cover the diverse applications of nanotechnology in sustainable finishing, focusing on textiles, agriculture, construction, and other key sectors. This will also examine the associated challenges and consider future directions for this promising field.

Nanotechnology, the manipulation of matter at the atomic and molecular level, offers a revolutionary approach to problem-solving across diverse fields. Its promise lies in the unique properties that materials exhibit at the nanoscale, which can be vastly different from their bulk counterparts. This opens up a world of possibilities for creating innovative solutions with enhanced performance, efficiency, and sustainability.

6 Nanoscale

The nanoscale refers to dimensions between 1 and 100 nm (a nanometre is one billionth of a metre). At this scale, materials can exhibit unique physical, chemical, and biological properties due to:

- **Increased Surface Area-to-Volume Ratio:** Nanoparticles have a much larger surface area compared to their volume, leading to increased reactivity and interaction with other materials.
- **Quantum Effects:** At the nanoscale, quantum mechanics plays a significant role, influencing the behaviour of electrons and leading to unique optical and electronic properties.
- **Surface Dominance:** The properties of nanomaterials are often dominated by their surface atoms, which have different bonding and interactions compared to atoms in the bulk material.

7 Unique Properties of Nanomaterials

These unique properties of nanomaterials can be harnessed to create a wide range of applications, including:

- **Improved Strength and Durability:** Nanomaterials can be used to create materials with exceptional strength, hardness, and durability, leading to lighter and longer-lasting products.
- **Efficient Conductivity:** Nanomaterials can exhibit excellent electrical and thermal conductivity, enabling the development of more efficient electronic devices and energy storage systems.
- **Catalytic Activity:** Nanoparticles can act as catalysts, accelerating chemical reactions and improving the efficiency of industrial processes.
- **Drugs Delivery:** Nanoparticles can be used to deliver drugs and other therapeutic agents to specific targets in the body, improving the effectiveness of treatments and reducing side effects.
- **Sensing and Detection:** Nanosensors can detect minute quantities of chemicals, pollutants, or biological molecules, enabling early diagnosis of diseases and environmental monitoring.

8 Application of Nanotechnology across Industries

The promise of nanoscale solutions is being realized across various industries:

- **Medicine:** Nanotechnology is revolutionizing drug delivery, diagnostics, and regenerative medicine, leading to more effective treatments for diseases like cancer and Alzheimer's.
- **Electronics:** Nanomaterials are enabling the development of faster, smaller, and more energy-efficient electronic devices, including smartphones, computers, and displays.
- **Energy:** Nanotechnology is being used to develop more efficient solar cells, batteries, and fuel cells, contributing to the development of renewable energy technologies.
- **Environmental Science:** Nanomaterials are being used for water purification, pollution remediation, and the development of sustainable agricultural practices.
- **Manufacturing:** Nanotechnology is being used to create stronger, lighter, and more durable materials for various applications, including automobiles, aircraft, and construction.

While the promise of nanoscale solutions is immense, it is important to address the potential risks associated with nanotechnology. Careful research and development are needed to ensure the safety of nanomaterials and their responsible use. By addressing these challenges and fostering collaboration between researchers, industry, and policymakers, we can unlock the full potential of nanotechnology to create a more sustainable and prosperous future [25].

9 Application of Nanotechnology in Textiles

9.1 Hydrophobic Nanofinish in Textiles

The two primary methods for producing hydrophobic surfaces are (i) forming a rough structure on a hydrophobic surface and (ii) altering a rough surface with low surface free energy materials. Both methods have been used to textile substrates to give them a hydrophobic finish.

The fabric's ability to resist water is achieved by adding nanowhiskers, which are fluorocarbons and 1/1000 the size of a normal cotton fibre, to the fabric to provide a peach fuzz effect without weakening the cotton's strength. As a result, a rough hydrophobic layer forms. A family of organic compounds known as fluorocarbons has perfluroalkyl residues, where fluorine has taken the place of a hydrogen atom. These substances are little reactive and have very good thermal stability.

9.2 Super Hydrophobic of Textile Material

As previously mentioned, hydrophobic fluorocarbon coatings reduce surface energy and can provide a maximum water contact angle of about 120. A superhydrophobic finish with a contact angle greater than 160 is necessary to achieve a higher contact angle and self-cleaning capabilities. Surface coating is not an option for this kind of finish. A big geometric area for a comparatively modest projected area is provided by a super hydrophobic increase in surface roughness. A substrate component with a variety of microscale to nanoscale projections or voids is typically the shape of the roughened surface. Water and air that was trapped between the surface's cracks caused the rough surface to repel. This minimizes the solid/water interaction while increasing the air/water interface. Spreading doesn't happen here; instead, the water forms a spherical droplet.

9.3 Self-Cleaning Effect in Textile Materials

The ability of tough plant leaves to self-clean was examined. The majority of the 340 plant species that were examined had wettable leaves that were relatively smooth and devoid of noticeable surface sculptures (<110). On the other hand, water-repellent leaves have various surface sculptures, primarily papillose epidermal cells, and epicuticular wax crystal. They have more than 160. They noticed that water condensed to form spherical droplets on water-repellent surfaces. Even at a minor angle of inclination (less than five), it easily came off the leaf without leaving any trace.

When the leaf was exposed to either artificial or natural rain, all particles that were stuck to the leaf surface were completely eliminated. Because the dirt deposited on the waxy surface of the leaves is often greater than the microstructure of the leaf's surface, it is deposited on the tips, minimizing the interfacial area between the two. The surface area of the water droplet exposed to air is decreased and energy is acquired by absorption in this instance of the drop rolling over a particle. The particle is taken up by the water drop and extracted from the leaf surface as a result of the particle and water droplet sticking. The lotus effect is the name given to this phenomenon. Superhydrophobicity, a very high water contact angle, and a very low roll-off angle are the two conditions that determine the lotus effect.

The Nanosphere was created by the Swiss textile manufacturer Schoeller's Nano-Tex to create water-repellent textiles. A three-dimensional surface structure with gel-forming chemicals that resist water and stop dirt particles from sticking is used in Nanosphere impregnation. The rough, textured surfaces of lotus plants are extremely hydrophobic. When water droplets land on them, they bead up and roll off if the surface slopes slightly. Because of this, even after a long shower, the surfaces remain dry. Additionally, even with little rain, the lotus plant's leaves remain clean because the raindrops collect tiny dirt particles as they roll [26].

9.4 UV Protection of Textiles

In the past, textile materials were covered with organic UV absorbers, which successfully blocked UV rays but were not as long-lasting. Typically, semiconductor oxides like TiO2, ZnO, SiO2, and Al2O3 are UV blockers. Zinc oxide (ZnO) and titanium dioxide (TiO2) are two of the most often utilized semiconductor oxides. It was shown that zinc oxide and titanium dioxide nanoparticles were more effective in absorbing and dispersing UV light than their conventional counterparts, making them more effective UV blockers. This is because nanoparticles are more effective at blocking UV light than traditional materials since they have a higher surface area per unit mass and volume. Light scattering dominates at around a tenth of the scattered light's wavelength for tiny particles.

According to Raleigh's scattering hypothesis, the scattering was inversely proportional to the wavelength to the fourth power, meaning that it was highly reliant on the wavelength. According to this hypothesis, the ideal particle size will be between 20 and 40 nm in order to scatter UV light between 200 and 400 nm. Sol-gel was used to create a UV-blocking coating for cotton textiles. Excellent UV protection is provided by the thin coating of titanium dioxide that forms on the surface of the treated cotton fabric; this effect lasts for up to 50 home washings. In addition to titanium dioxide and water droplet rolls, cotton fabric was treated with zinc oxide nanorods ranging in length from 10 to 50 nm to offer UV protection. The fabric coated with zinc oxide nanorods showed a good UV protection factor (UPF) rating, according to the investigation of the UV-blocking effect.

9.5 Antibacterial Finish in Textiles

Textile fibres, whether synthetic or natural, are susceptible to harmful fungus and bacteria. As a result, methods for antibacterial disinfection and finishing have been developed for a variety of textile types. One such method is the application of nano-sized silver, titanium dioxide, and zinc oxide colloidal solutions (25–50 ppm) to cotton and polyester fabrics to treat the fibres. A certain level of sterilizing impact is exhibited by metallic ions and metallic compounds. It is believed that photocatalysis with the metallic ion converts a portion of the oxygen in the air or water into active oxygen, dissolving the organic material and producing a sterilizing effect. Because there are more particles per unit area when nanoparticles are used, the antibacterial effects can be enhanced.

The bactericidal and fungicidal efficacy of nanosilver particles is greatly enhanced by their incredibly huge relative surface area, which increases their interaction with bacteria or fungi. Proteins react strongly with nanosilver. It will negatively impact cellular metabolism and impede cell development when it comes into touch with bacteria and fungi. Additionally, it inhibits the transport of the substrate into the microbial cell membrane, respiration, and the electron transfer system's basal metabolism. Additionally, it prevents the growth and proliferation of the bacteria and fungi that cause infections, odours, itching, and sores. As a result, socks are frequently treated with nanosilver particles to stop germs from growing. In addition, nanosilver can be applied to a range of other healthcare products such as dressings for burns, scald, skin donor and recipient sites.

Inks and dyes that include concentrated concentrations of fast-drying colourants are known as Nano-Tex Resists Spills Fabric Protection Repel Ink. Although NANO-TEX Resists Spills protection does not shield fabrics from certain colourants, stains can be reduced if the fabric is removed promptly.

The fabric protection offered by NANO-TEX Resists Spills guards against liquids that are both water- and oil-based. Superior liquid repellency provided by Resists Spills fabric protection reduces stains. The wearer may carefully lift off the dirt if it is sitting on the fabric's surface thanks to NANO-TEX fabric protection, which reduces discolouration. To reduce or eliminate the possibility of stains, the suggested washing instructions for garments should be adhered to if the mud has been ground into the NANO-TEX Resists Spills enhanced fabric. Some discolouration may remain when mud stains or strong oils are rubbed over textiles.

9.6 Photocatalysis of Nanofinish

Ultraband gap light transforms semiconductor materials into potent redox catalysts that can clean water, kill germs, and even split water into hydrogen and oxygen.

When exposed to light with energy greater than its band gaps, the electrons in titanium dioxide (TiO2) will move from the valence band to the conduction band, forming electron (e-) and electric hole (h) pairs on the photocatalyst's surface. Oxygen and negative electrons will unite to form O2, whereas water and positive electric holes will produce hydroxyl radicals. Both are unstable chemicals, thus when the organic molecule hits the photocatalyst's surface, it will react with O2 - and OH -, respectively, to form water (H2O) and carbon dioxide (CO2) [27].

The photocatalyst may break down typical organic compounds in the air, including bacteria, viruses, and smell molecules, through the process. Because of its photocatalytic activity, it was found that a cloth coated with nano-TiO2 may effectively defend against germs and stain discolouration. However, zinc oxide also functions as a photocatalyst, and its process of photocatalysis is comparable to that of titanium dioxide; the only difference is the band gap. When exposed to light, nano-ZnO has potent photocatalytic capabilities, which is why it is used to provide textiles antibacterial qualities.

9.7 Nanocare of Textile Material

For use with cellulosic fibres like cotton and linen, this technique creates a completely carefree fabric with water and stain repellent, shrink-proof, and wrinkle-resistant qualities. It is a next-generation, dimension-stabilizing, easy-to-maintain finish that goes beyond techniques that only provide shrink-proofing and wrinkle resistance. Nanocare can survive over 50 household washings. Even after 20 home washes, it retains high water and oil repellence levels (80 and 4), imparting stain resistance and water repellence that are superior to those of traditional procedures. The important properties include Water, Oil, and Stain Repellence and Preserves Original Hand Ease of Care Breathable Fabric, Wrinkle resistant.

In the nanopel, the application of the water-and-oil repellent finishing using nanotechnology works well on synthetic materials like polyester, nylon, and acrylic as well as natural fibres like cotton, linen, wool, and silk. It is possible to anticipate unparalleled durability and grease and water resistance, especially with natural fibres. Even after 20 washes, Nano-Pel cotton maintains its water and oil repellence (80 and 4) at a high level, withstanding 50 home launderings.

Nanodry is a hydrophilic finishing technique that gives exceptional durability over 50 home washings and presents opportunities for significant advancement in the field of synthetic clothing made of polyester and nylon. When deep-dyed, nanodry prevents dye migration and offers durability that is superior to that of the hydrophilic polyester finishing that is frequently done in Japan utilizing polyethylene glycol polymer molecules. Since there isn't a long-lasting hydrophilic coating for nylon, it is anticipated to work especially well for sportswear and knickers that need to absorb sweat. Within the next 3 to 6 months, significant growth is anticipated, mostly in the sportswear sector.

The nanotouch finish gives the durable cellulose wrapping over synthetic fibre is provided by this cutting-edge finishing method. Together, the cellulosic sheath and synthetic core create a concentric structure that addresses the drawbacks of synthetics, including their hydrophobicity, electrostatic properties, artificial hand, and blinding lustre. Without the drawbacks of synthetic suits, such as their hydrophobic, electrostatic, and unnatural hand, it will expand the current use of synthetics. Examples of novel applications made possible by Nano-Touch, a new fibre compounding standard, include the following.

Traditional textile coating is challenged by self-assembled nanolayer (SAN) coating. This field of study is still in its infancy. Target chemical molecules create a layer on the surface of textile materials that is less than a nanometre thick in self-assembled nanolayer (SAN) coating. A nanolayered structure can be created by stacking more layers on top of the preexisting ones. To provide textile materials unique functionality, several SAN techniques are being investigated [28].

10 SAN Coatings and Applications

- Superior refinement in a blended fabric.
- Durable performance.
- Luxurious cotton-like hand.
- Easy care.
- Reduced static build-up.

11 Nanomaterials Used for Different Textiles Fabrics and Properties

Nanofibres have amazing potential and are also used in the fabric industry. Research on using nanotechnology to improve functionality or create previously unheard-of textile fabric processes is flourishing. It is quite likely that nanotechnology will permeate every aspect of textiles in the years to come [29].

Since the law mandates that the industry utilize low flammability materials, ongoing research is being done to produce flame-retardant products since synthetic fibres like polyester and polyamide are extremely flammable [30].

The use of nanoparticles as a flame retardant in finishing textiles has recently emerged as a viable and intriguing strategy. Although not all of the nanoparticles on the market can be utilized in the flame retardancy sector, the primary benefit is that a little quantity of them may be used [31].

Silver stands out among all metallic minerals due to its potent antibacterial properties, which demonstrate antimicrobial activity against bacteria, viruses, and microorganisms [32]. One significant feature is that silver exhibits good biocompatibility and minimal toxicity to human cells in both ionic and colloidal forms [33].

The scientific community is considering a number of theories, but the true mechanism of silver's antibacterial effect is still unclear. While some research focus on the negative effects of silver nanoparticles (Ag NPs) in the cell membrane, claiming that these Ag NPs cause cell death, other studies show that the antibacterial impact is caused by the release of silver ions from Ag NPs. These methods are described in more depth in [34].

Nanowhiskers are hydrocarbons that are a thousandth of the size of a typical cotton fibre that is attached to the cloth.

Although they are smaller than typical water droplets, the gaps inside cotton whiskers are more significant than water molecules. Water thus remains on the outside of the fabric and on the hairs or whiskers, but it may now pass through the cotton material when pressure is applied [35].

Structural polymers that are prone to cracking and other forms of damage are called composite materials. The structure is negatively impacted by these fissures, and it is practically impossible to discover and fix them. The stability of the material is seriously jeopardized if a crack appears within the composite. Fabric support and a fastening matrix, usually composed of a polymeric material, combine to form fabric composites. Fabric compounds are mostly used in the fields of building, power production, land transportation, marine, aerospace, and defence.

12 Textiles and Nanoparticles

These consist of clay, carbon black, and metal oxides. In addition to biological and chemical varieties, metal oxides at the nanoscale are inherently photocatalytic, UV absorptive, electrically conductive, and photo-oxidizing. Examples of metal oxides include ZnO, TiO2, MgO, and Al2O3. In order to create complex fibres that are antistatic and block UV rays, ZnO is added to nylon while making swimwear fabric. Clay nanoparticles may block UV rays and are resistant to heat, electricity, and chemicals.

Thus, adding it to fabric can result in improved modulus, tensile strength, and flexural strength. Clay nanoparticle-based nanocomposite fibres have the potential to be corrosion-, flame-, and UV light-resistant. In order to make nylon flame resistant without releasing deadly gases, clay nanoparticles are incorporated into the fabric. Composite fibres' tensile strength may be effectively increased by adding carbon nanofibre and nanoparticles.

Graphite nanofibres (GNF) and carbon black nanoparticles are two commonly used nanofillers in the fabric industry that improve the durability and abrasion resistance of composite fibres. GNFs and other complex fibres have higher tensile strength. Matrix materials such as polyester, nylon, and polyethylene have been used for these nanofillers [36].

Photo-catalysts include titanium dioxide (TiO2). Because of the photocatalytic effect, fabric treated with nano-TiO2 can provide a useful barrier against germs and staining colour. The TiO2 electron is illuminated by light with larger energy gaps in the band [37].

Given the unique benefits of employing nanostructured materials in the textile sector, this section focusses on altering the surface of textiles using various nanostructured materials that exhibit superior antibacterial qualities. Silver stands out among all metallic minerals due to its potent antibacterial properties, which demonstrate antimicrobial efficiency against bacteria, viruses, and germs.

In situ production of Ag NPs into the electrospinning solution is the basis of the second technique. Silver nitrate (AgNO3) is added to the polymeric solution for this reason. In this case, the polymer serves as the silver cations' encapsulating agent. Ag NPs are then created by adding a reducing agent to the solution. Sometimes the solvent itself serves as a reducing agent. Examples of constructed electrospun carbon nanofibers (ECNFs).

Given that UV finishing is currently one of the key characteristics of textile materials, a great deal of study has been done in this area. Using the sol-gel process, UV blocking treatments for cotton textiles are created. Generally speaking, there are two types of UV blockers: inorganic and organic. Because they are non-toxic and chemically stable under UV and high temperatures, inorganic UV blockers are preferred [38].

13 Nanoribbons in Textiles

Nanoribbons are the most popular type of electrical circuit. In terms of flexibility, these are very little electronic sheets. These ribbons serve as the foundation for chips with a high degree of stretchability and elasticity. Researchers are concentrating on creating and using these ribbons to either line the brain to track activity in those who may be at risk for epilepsy or include them into surgical gloves to track a patient's vital signs as they undergo surgery.

Textile-based nanosensors that can sense their environment and communicate with the wearer have been made possible via nanotechnology. When the body is active, these sensors can track our vital signs and react to weather changes. Sports items that have benefited from nanotechnology include skin treatments, tennis racquets, golf balls, and running shoes.

Even synthetic fibres, which often have weak antistatic qualities, can benefit from nanotechnology's antistatic capabilities. According to studies on antistatic finishing, synthetic fibres might be given antistatic qualities by nanoparticles such as silanenanosol, zinc oxide whiskers, titanium dioxide, and nanoantimony-doped tin oxide (ATO). These substances aid in removing the fabric's build up of static charges. The nanoparticles become permanently bonded to the fibres when the fabric is coated with an antistatic treatment, forming an electrically conductive network. The isolated charged regions and voltage peaks that are often present in antistatic materials are prevented by this conductive network.When zinc oxide finishing is completed, cotton fabric is treated with nanorods composed of zinc oxide, which are typically 10–50 nm long. According to studies on Nanofinishes, cloth treated with zinc oxide Nanorods showed an exceptional UV protection factor (UPF) rating [39].

Additionally, this method prevents the growth and proliferation of the bacteria and fungi that cause infections, odours, itching, and sores [38].

Since AuNPs are the most stable metal nanoparticles, various applications are made possible by their easy preparation and oxidation resistance. AuNPs have notable catalytic activity across a range of reaction systems. They have a high surface-to-volume ratio, exceptional biocompatibility with the right ligands, and distinctive optoelectronic characteristics. Through manipulation of their size, shape, and environment, AuNPs' characteristics, including the LSPR optical effect, may be easily adjusted.

Furthermore, AuNPs have been extensively employed as efficient active substrate materials for SERS investigation. It is possible to significantly improve the Raman signals of molecules adsorbed on AuNPs. Some of these crucial characteristics can be transferred to the final textile goods by the modification of textile materials using AuNPs. The most common natural fibre material used to make textiles and apparel is cotton. Growing customer desire for more functional traditional textile items is driving cotton modification. A straightforward and efficient method for functionally altering textile materials is in situ synthesis of nanoparticles on fabric or fibres. For instance, in situ development of silica nanoparticles has produced a hydrophobic hierarchical structure on cotton fabric [40].

With a diameter as tiny as a nanometre (nm), carbon nanotubes are tubular forms of carbon. A two-dimensional graphene sheet coiled into a tube is configurationally comparable to a carbon nanotube (CNT). Depending on their chirality, they may be semiconducting or metallic.

Because of their great electrical conductivity and strength, carbon nanotubes (CNTs) are among the most promising materials. CNT is made up of one or more microscopic graphite shells that have been folded into cylinders [41].

At one-sixth of the weight of steel, carbon nanotubes (CNTs) have 100 times the tensile strength, a thermal conductivity superior to all but the finest diamond, and an electrical conductivity comparable to copper but capable of carrying far larger currents. CNTs have the potential to be used in sensors, field emission displays, energy storage and conversion devices, conductive and high-strength composite fibres, and more. Screen displays, sensors, aeroplane construction, explosion-proof blankets, and electromagnetic shielding are a few examples of potential uses. The composite fibres may find use in electromagnetic shielding, safety harnesses, and blankets that are resistant to explosions. The investigation of various fibre polymer matrices, including polymethylmethacrylate (PMMA) and polyacrylonitrile (PNA), as well as CNT dispersion and orientation in polymers, are all part of ongoing research on CNT fibres [42].

Synthetic fibres may acquire antistatic qualities from nanoparticles such as silanenanosol, zinc oxide whiskers, titanium dioxide, and nanoantimony doped tin oxide (ATO). This type of material aids in efficiently dissipating the static charge that builds up on the cloth. However, silanenanosol enhances antistatic qualities because the fibre's silica gel particles absorb moisture and water from the air through the use of bound water and amino and hydroxyl groups. Durably fixed in the teflon membrane's fibrils, electrically conductive nanoparticles form an electrically conductive network that inhibits the development of isolated charged regions and voltage peaks that are frequently seen in traditional antistatic materials.

When applied to fabric, nanowhiskers—hydrocarbons that are 1/1000 of the size of a conventional cotton fibre—create a peach fuzz look while maintaining the fabric's water-repellent qualities without weakening the cotton's strength. Water stays on top of the whiskers and above the fabric's surface because the gaps between the whiskers on the fabric are less than a normal drop of water but yet larger than water molecules. Nevertheless, if pressure is put on the cloth, liquid can still flow through it.

14 Nanotubes in Textiles

Textile fibres enhanced with carbon nanotubes (CNTs) have been produced to reduce weight and increase strength and toughness. Polypropylene and carbon particles might be melt spun to create these polymeric textile fibres. It has been demonstrated that adding CNTs to fibres increases their strength and functionality. Along the weft and warp directions, the CNT-coated cotton textiles' tensile strength increased, demonstrating improvements in loading capacity and flexibility [43].

Traditionally, to give textiles wrinkle resistance, resin is infused into the cloth. NPs have been added to polymeric textile fibres to provide them wrinkle resistance [44]. Excellent mechanical strength, antistatic, antibacterial, and UV absorption qualities are among the many outstanding qualities of ZnO nanomaterials. The UV visible spectrum of ZnO nanoparticles synthesized with 0%–5% soluble starch reveals peak absorption at 361 nm. 3·443 eV is the energy determined using Planck's constant. 40 nm was determined to be the size (diameter) using the effective mass approximation.

A key component of textile materials' performance that serves as a gauge for coated textiles' breathability is air permeability. Although applying a bulk ZnO coating to cloth decreases air permeability, applying a nano-ZnO coating greatly increases it. As a result, the breathability of the air was enhanced by the even and extremely thin distribution of nano-ZnO. There was no discernible change in bursting strength as a result of coating [45].

15 Conductive Metals in Fabrics

Metals, such as copper, silver, and steel, are the most electroconductive materials and are extensively used in smart textile applications such as copper yarns and silver-coated fibres. Silver (Ag), gold (Au), copper (Cu), titanium dioxide (TiO2), zinc oxide (ZnO), nanolayered clay, and other organic and inorganic nanoparticles and their nanocomposites are applied to textile materials to pass on different multifunctional properties. The conductivity of metallic-nanoparticle-coated threads is low because their small size leads to many interparticle junctions. A few metals and metal-derived oxides used in textile finishing are discussed here.

16 Silver Nanoparticles (AgNPs) Finish in Textiles

AgNPs are known to be antibacterial, and their therapeutic properties have been proven against a broad range of microorganisms that can cause diseases. Ag can be synthesized through various methods ranging from simple chemical reduction processes to photochemical, biochemical, so no chemical, and Tollens' reagent methods. The deposition of nanowire films around threads can be achieved through dip-coating. Chemically metallized Ag threads are used to produce textiles with antistatic properties or to shield against electromagnetic waves. Ag-coated fibres are used to improve the behaviour of smart textiles.

17 Gold Nanoparticles (AuNPs) Finish in Textiles

AuNPs are widely known for their optical and biological properties. They could be used for highly sensitive diagnostic assays, thermal ablation, radiotherapy enhancement, and drug and gene delivery. They are mainly prepared via the chemical reduction method using gold salts such as hydrogen tetrachloroaurate (HAuCl4) as a precursor and citrate as a reducing agent. Au nanoparticles chemically synthesized by citrate reduction can be immobilized onto chitosan-treated soybean knitted fabric via the exhaustion method. A gold coating with solution deposition on knitted fabric is highly conductive, stretchable, wearable, and washable.

18 Copper Nanoparticles (CuNPs) in Textiles

CuNPs have heat transfer properties such as high thermal conductivity, low preparation cost, antibacterial efficiencies, catalytic activity, and electrical and optical properties. They are produced by the chemical reduction method. The surface of the textile fabric can be covered with a layer of densely packed Cu particles after ELD. Cu-coated textiles can show a bactericidal effect. The antibacterial property is important because electromagnetic interference (EMI) shielding fabrics are excellent media for microorganism growth.

19 Metal Oxide Nanoparticles (MONPs) in Textiles

MONPs are renowned for environmental remediation and electronic applications. The particles can generate charge carriers through energy absorption. TiO2 and ZnO are widely used in textile finishing. Nanoiron oxide, nanocopper oxide, nanocobalt oxide, and nanomanganese oxide in cotton fabrics can be created for high antibacterial efficiency. In addition, UV protection and flame retardation can be developed by in situ preparation of nanometal oxides [46].

20 Conclusion

New research on the application of nanotechnology has sparked concerns about the absence of industrial regulation, safety assessments, and health data about possible effects on the environment and human health. There is enough evidence from early nanotechnology research to suggest that nanoparticles may have unique toxic characteristics. According to others, the very characteristics that make nanomaterials appealing to the market might also make them more hazardous than their equivalents of regular size.

"Nanowhiskers," or nanofibres, may contain harmful substances like fluorotelomers. International organizations are demanding adequate and appropriate regulation, safety testing, and assessment of the developing field of nanotechnology, particularly those nanomaterials currently used extensively in commerce.

Bringing nanoscale manufacturing from a lab to an industrial setting is a challenging undertaking. The negative impacts of nanoparticles in commercial manufacturing have been the subject of very few studies. Potential explosions and the potential for nanomaterials to catch fire in the presence of air if hazardous dangers are not properly managed. A number of scientists have expressed concern about potential contamination. Layer-free nanomaterials have the potential to become a new kind of pollution. Both bacteria and animals may be at risk from nanomaterials.

Although there are benefits to using nanotechnology into textiles, there are also safety and health risks. When breathed or consumed, nanoparticles produced from textiles during production, usage, and disposal may have toxicological consequences. To comprehend these hazards, evaluate toxicity, and create safety regulations, a great deal of study is being done. Regulations and occupational exposure limits tailored to the textile industry's use of nanoparticles are being established.

Responsible waste management techniques and sustainable production methods should be used, and the effects on the environment must be taken into account. To guarantee the safe application of nanotechnology in textiles, extensive research on the toxicity and long-term impacts of nanoparticles is required. It is crucial to take a preventative strategy that include risk evaluations, exposure restrictions, and training. For the safe development of textiles improved with nanomaterials, cooperation between academic institutions, industry, and regulatory agencies is essential.

References

1. Santamaria, A. (2012). *Nanotoxicity* (pp. 1–12). Humana Press.
2. Marques, A. C., Vale, M., Vicente, D., Schreck, M., Tervoort, E., & Niederberger, M. (2021). *Global Challenges, 5*, 2000116.
3. Sharifi, S., Behzadi, S., Laurent, S., Laird Forrest, M., Stroeve, P., & Mahmoudi, M. (2012). *Chemical Society Reviews, 41*, 2323–2343.

4. Wong, Y., Yuen, C., Leung, M., Ku, S., & Lam, H. (2006). Selected applications of nanotechnology in textiles. *Autex Research Journal, 6*(1), 1–10.
5. Shabanloo, R., Karimi, N., & Montazer, M. (2024). Chapter 13: Nanomaterials and textile industry. In M. I. Malik, D. Hussain, M. R. Shah, & D.-S. Guo (Eds.), *Micro and nano technologies, Handbook of nanomaterials* (Vol. 1, pp. 309–337., ISBN 9780323955119). Elsevier. https://doi.org/10.1016/B978-0-323-95511-9.00002-0
6. Dolez, P. I. (2019). Chapter 6—Application of nanomaterials in textile coatings and finishes. In P. N. Tri, S. Rtimi, & C. M. Ouellet Plamondon (Eds.), *Micro and nano technologies, Nanomaterials-based coatings* (pp. 139–169., ISBN 9780128158845). Elsevier. https://doi.org/10.1016/B978-0-12-815884-5.00006-5
7. ww.luxresearchinc.com, 2013.
8. Cramer, D. R., Ponomarenko, A. E., Laurent, S., & Burckett, J. C. T. R. (2003). U.S. Pat. No: 6 645. p. 569.
9. Mishra, R., Militky, J., Baheti, V., Huang, J., Kale, B., Venkataraman, M., et al. (2014). The production, characterization and applications of nanoparticles in the textile industry. *Textile Progress, 46*(2), 133–226. https://doi.org/10.1080/00405167.2014.96441
10. Sailor, M. J., & Lee, E. J. (1997). *Advanced Materials, 9*, 783.
11. Prasad, S. R., Kumbhar, V. B., & Prasad, N. R. (2023). Applications of nanotechnology in textile: A review. *ES Food and Agroforestry*, 1019/1-9. https://doi.org/10.30919/esfaf1019
12. Singh, M., Vajpayee, M., & Ledwani, L. (2020). Eco-friendly surface modification and nanofinishing of textile polymers to enhance functionalisation. In *Nanotechnology for energy and environmental engineering* (pp. 529–559). Springer.
13. Voelker, D., Schlich, K., Hohndorf, L., Koch, W., Kuehnen, U., Polleichtner, C., Kussatz, C., & Hund-Rinke, K. (2015). Approach on environmental risk assessment of nanosilver released from textiles. *Environmental Research, 140*, 661–672.
14. Quadros, M. E., Pierson, R., Tulve, N. S., Willis, R., Rogers, K., Thomas, T. A., et al. (2013). Release of silver from nanotechnology-based consumer products for children. *Environmental Science & Technology, 47*(15), 8894–8890.
15. Koncar, V. (2016). Introduction to smart textiles and their applications. In *Smart textiles and their applications* (pp. 1–8). Elsevier Inc.
16. Kumar, R.; Chauhan, M.; Sharma, N.; Chaudhary, G.R2018. Toxic effects of nanomaterials on environment. In Environmental toxicity of nanomaterials; CRC Press, ,; pp. 1–20.
17. Johnston, L. J., Gonzalez-Rojano, N., Wilkinson, K. J., & Xing, B. (2020). Key challenges for evaluation of the safety of engineered nanomaterials. *Nano Impact, 18*, 100219.
18. Kawai, S., Niwano, M., & Sato, M. (2019). A risk assessment framework for self-management of poorly soluble low toxic nanomaterials. *Heliyon, 5*, e02165.
19. Aryal, S., Park, H., Leary, J. F., & Key, J. (2019). Top-down fabrication-based nano/microparticles for molecular imaging and drug delivery. *International Journal of Nanomedicine, 14*, 6631.
20. Sukhanova, A., Bozrova, S., Sokolov, P., Berestovoy, M., Karaulov, A., & Nabiev, I. (2018). Dependence of nanoparticle toxicity on their physical and chemical properties. *Nanoscale Research Letters, 13*, 44.
21. Ramachandran, M. S. (2014). *Heart and toxins*. Elsevier.
22. Saleem, H., & Zaidi, S. J. (2020). Sustainable use of nanomaterials in textiles and their environmental impact. *Materials, 13*(22), 5134. https://doi.org/10.3390/ma13225134
23. Harifi, T., & Montazer, M. Application of nanotechnology in sports clothing and flooring for enhanced sport activities, performance, efficiency and comfort: A review. *Journal of Industrial Textiles, 46*, 1147–1169.
24. Almeida, L., & Ramos, D. (2017). Health and safety concerns of textiles with nanomaterials. *IOP Conference Series: Materials Science and Engineering, 254*, 102002.
25. Russell, E. (2002). Nanotechnologies and the shrinking world of textiles. *Textile Horizons, 9*(10), 7–9.

26. Cramer, Dean, R. Ponomarenko, Anatolyevna E., Laurent, S., and Burckett, J.C.T.R., Method of applying nanoparticles. U.S. Pat. No: 6,645,569, 2003.
27. Anonymous. (2003). Small-scale technology with the promise of big rewards. *Technical Textiles International, 3*, 13–15.
28. Xin, J. H., Daoud, W. A., & Kong, Y. Y. (2004). A new approach to UV-blocking treatment for cotton fabrics. *Textile Research Journal, 74*, 97–100.
29. Haque, M. (2019). Nano fabrics in the 21st century: A review. *Asian Journal of Nanoscience and Materials, 2*(2), 143.
30. Levchik, S. V., & Weil, E. D. (2005). Flame retardancy of thermoplastic polyester—A review of the recent literature. *Polymer International, 54*, 11–35.
31. Alongi, J., Tata, J., Carosio, F., Rosace, G., Frache, A., & Camino, G. (2015). A comparative analysis of nanoparticle adsorption as fire-protection approach for fabrics. *Polymers, 7*:47–68; Carosio, F., Alongi, J., Frache, A. (2011). Influence of surface activation by plasma and nanoparticle adsorption on the morphology, thermal stability and combustion behavior of PET fabrics. *European Polymer Journal, 47*:893–902.
32. Morones, J. R., Elechiguerra, J. L., Camacho, A., Holt, K., Kouri, J. B., Ramírez, J. T., & Yacaman, M. J. (2005). The bactericidal effect of silver nanoparticles. *Nanotechnology, 16*:2346–2353; Panácek, A., Kvítek, L., Prucek, R., Kolár, M., Vecerová, R., Pizúrová, N., Sharma, V. K., Nevecná, T., & Zboril, R. (2006). Silver colloid nanoparticles: Synthesis, characterization, and their antibacterial activity. *The Journal of Physical Chemistry. B 110*: 16248–16253.
33. Marambio-Jones, C., & Hoek, E. M. V. (2010). A review of the antibacterial effects of silver nanomaterials and potential implications for human health and the environment. *Journal of Nanoparticle Research, 12*, 1531–1551.
34. Radetic, M. (2012). Functionalization of textile materials with silver nanoparticles. *Journal of Materials Science, 48*, 95–107.
35. Nakamura, J., Nakajima, N., Matsumura, K., & Hyon, S. H. (2011). *Journal of Nanomedicine & Nanotechnology*.
36. Gosh, S. K. (2006). *Functional coatings by polymer microencapsulation* (1st ed.). Wiley-VCH Verlag GmbH & KG.
37. Pandurangappa, C., & Lakshminarasappa, B. N. (2011). *Journal of Nanomedicine & Nanotechnology, 2*:108; Patil, A., Chirmade, U. N., Trivedi, V., Lamprou, D. A., Urquhart, A., & Douroumis, D. (2011). *Journal of Nanomedicine & Nanotechnology*, *2*(3):111; Adjah AD (2011) Hydrol Current Res. 2: 116.
38. Yang, H. Y., Zhu, S. K., & Pan, N. (2003). Studying the mechanisms of titanium dioxide as ultraviolet-blocking additive for films and fabrics by an improved scheme. *Journal of Applied Polymer Science, 92*, 3201–3210.
39. Wong, Y. W. H., Yuen, C. W. M., Leung, M. Y. S., Ku, S. K. A., & Lam, H. L. I. (2006). Selected applications of nanotechnology in textiles. *AUTEX Research Journal, 6*(1), 1–8.
40. Chen, X. Q., Liu, Y. Y., Lu, H. F., Yang, H. R., Zhou, X. A., & Xin, J. H. (2010). In-situ growth of silica nanoparticles on cellulose and application of hierarchical structure in biomimetic hydrophobicity. *Cellulose, 17*: 1103–1113; Sadr, F. A., Montazer, M. (2014). In situ sonosynthesis of nano TiO2 on cotton fabric. *UltrasonSonochem, 21*: 681–691.
41. Wang, C. C., & Chen, C. C. (2005). Physical properties of cross-linked cellulose catalyzed with nano titanium dioxide. *Journal of Applied Polymer Science, 97*, 2450–2456.
42. Scott, M., & Holly, H. (2007). *Axtell, the next generation of chemical and biological protective material utilizing reactive nanoparticles*. Gentex Corporation.
43. Liu, Y., Wang, X., Qi, K., & Xin, J. (2008). Functionalization of cotton with carbon nanotubes. *Journal of Materials Chemistry, 18*, 3454–3460.
44. Yu, M., Gu, G., Meng, W. D., & Qing, F. L. (2007). Superhydrophobic cotton fabric coating based on a complex layer of silica nanoparticles and perfluorooctylated quaternary ammonium silane coupling agent. *Applied Surface Science, 253*, 3669–3673.

45. Yadav, A., Prasad, V., kathee, A. A., Raj, S., Yadav, D., Moorthy, C. S., & Vigneshwaran, N. (2006). Functional finishing in cotton fabrics using zinc oxide nanoparticles. *Bulletin of Materials Science, 29*(6), 641.
46. Rayhan, M. G. S., Khan, M. K. H., Shoily, M. T., Rahman, H., Rahman, M. R., Akon, M. T., Hoque, M., Khan, M. R., Rifat, T. R., Tisha, F. A., Sumon, I. H., Fahim, A. W., Uddin, M. A., & Sayem, A. S. M. (2022). Conductive textiles for signal sensing and technical applications. *Signals, 4*(1), 1–39. https://doi.org/10.3390/signals4010001

Index

S. S. Muthu (ed.), *Sustainable Nano Textile Technology*, Textile Science and Clothing Technology, https://doi.org/10.1007/978-3-032-05902-4